Reliability-based Modeling of System Performance

**Reliability of Multiphysical Systems Set**

coordinated by
Abdelkhalak El Hami

Volume 19

# Reliability-based Modeling of System Performance

Abdelkhalak El Hami
Mohamed Eid

iSTE

WILEY

First published 2023 in Great Britain and the United States by ISTE Ltd and John Wiley & Sons, Inc.



ISTE Ltd
27-37 St George's Road
London SW19 4EU
UK

www.iste.co.uk

John Wiley & Sons, Inc.
111 River Street
Hoboken, NJ 07030
USA

www.wiley.com



Library of Congress Control Number: 2023938839

British Library Cataloguing-in-Publication Data
A CIP record for this book is available from the British Library
ISBN 978-1-78630-835-1

# Contents

# Introduction

Probabilistic modeling of system performance is built primarily on a solid foundation of reliability theory. It can therefore be called "reliability-based modeling of system performance". Although it has organic links with applied statistics, combinatorial calculations and Boolean algebra, reliability-based modeling is a distinct discipline that is conceptually different from the disciplines cited above. Currently, reliability-based modeling covers a very wide spectrum of applications, such as system reliability engineering, functional analysis, operational safety, risk analysis and management, risk management and maintenance optimization, system degradation and aging, system failure monitoring, diagnosis and prognosis, performance analysis, mass-production and quality control, structural reliability and software reliability.

The focus of the book is restricted to the reliability, availability, maintenance, and safety (RAMS) modeling of complex systems, excluding passive systems (structures) and software.

The book is accessible to engineering students and engineers who have already acquired the elementary knowledge of probability theory, applied statistics, differential calculus, integrals, combinatorial calculus and Boolean algebra.

Having specified above the areas contributing to the reliability-based modeling of system's performance, the primary functional qualities concerned are the availability and the reliability of the system. These functional qualities are probabilistic in nature and statistically measurable.

As soon as the concepts of availability and reliability are well understood, other derived qualities arise and complete the functional characterization of the system, such as the system functional space and the transition rates between the states defining the system functional space and the probabilities of sojourning in the states. Our recourse to mathematics does only serve to support the paradigm of reliability-based modeling of system's performance practiced by system engineering.

Throughout the book, the complexity of systems modeling is gradually built up. We start by modeling the operation of the simplest system that engineers can conceive, which is the "binary-state system". We will call it the elementary component. Progressively, we proceed towards the functional modeling of the multistate system. In this approach, the multistate system is built from independent and coherent elementary components. Independence and functional coherence of the constituents are well defined in the book. However, neither dependency nor functional incoherence of constituents has been addressed in the book.

Whether the system is binary or multistate, the definition of failure always remains of a functional nature. The physical roots of a failure may only appear at the lowest levels of the functional analysis where the elementary component failure appears. The lowest level of any functional analysis is conventionally decreed by the engineers and/or by the analysts. The functional boundaries separating between components and systems are also conventional. A system in a specific step of the analysis can become a component in the next step and vice versa. Gradually, this allows the modeling of the increasing functional complexity up to the system of systems. The book explores this possibility through the functional binarization of the multistate systems. The binarization of the multistate systems ensures the conservation of the fundamental mathematical models that describe the availability and the reliability of every functional entity through the full scale of complexity. Consequently, the availability and the reliability of every functional entity are governed by the same fundamental system of differential and integral equations, regardless of the degree of its complexity. The functional complexity is finally determined by the number of states defining the system, the functional space structure and the logical relations structuring the system functionalities. For low and medium complexity systems, the availability and the reliability can be determined by some analytical or numerical models. For higher complexities, analysts may call for the numerical simulation using the Monte Carlo simulation or similar methods. The book gives insights into these methods through simple didactic applications. The book is structured in eight chapters.

In Chapter 1, we insist on the identification of the system through its functionalities. As mentioned above, a system is a functional entity. As such, "availability" designates the aptitude of the system to supply its required functionalities on demand. It does not refer to its physical condition or being. A physically sound system may not be available because it is poorly cooled though its cooling unit is also sound. Indeed, it may not provide the required cooling capacity because the system operating temperature is beyond the safe thermal operating limits. Admittedly, the confusion between functionality and the physical system arises at the most elementary stage of the successive segmentation of functions and sub-functions. At the lowest stage of the segmentation, the elementary system would be arbitrarily called a component. The analyst arbitrarily sets the functional threshold below which the component would arise, in the most suitable way for his/her analysis. This threshold could obviously vary from one analysis to another. This elementary functional entity embodied in a component will be described in the simplest way that the systems engineer can imagine. It would exclusively be either "available" or "unavailable". The component therefore operates in perpetual transitions between the availability state and the unavailability state, until the end of its operational life. The transitions are characterized by a failure rate and a repair rate. The elementary component has a binary operating pattern, which is characterized by its failure and repair rates. Thus, we introduce the concept of the elementary component. The mathematical modeling of the availability and the reliability of the elementary component is developed in detail in Chapter 1.

Chapter 2 deals with the mathematical modeling of the availability and the reliability of the multistate system using the elementary models developed in Chapter 1. The multistate system is built by the functional assembly of the elementary components under the conditions that the elementary components are independent and functionally coherent. The independence requires that the transitions of the elementary components do not mutually infer. The functional coherence requires that the transition of any of the elementary components to an unavailability state cannot increase the overall availability of the system and that the transition of any of the elementary components to an availability state cannot increase the overall unavailability of the system. That is to say, in other words, the functional degradation of an elementary component can only degrade the overall functional state of the system, and the functional upgrading of an elementary component can only upgrade the overall functional state of the system. Having established the principles of independence and functional coherence, we only need to describe the logical functional relationships that structure all the elementary components within a well-defined system. These logical functional relations operate in two elementary algebraic modes, which are the union (OR) and the intersection (AND), in the Boolean sense of the terms. Then, the functional assembly of the elementary components in a single system is carried out using the elementary Boolean operators: OR "∨" and AND "∧"[1]. To introduce the notion of multistate system, we start by modeling the simplest multistate system one may conceive. That is a multistate system built up by only two elementary components. Some other multistate systems of higher order will equally be modeled before establishing the generic demarch of multistate systems modeling. Finally, we also demonstrate that any multistate system can be reduced to a binary elementary component by dividing its global functional space into a (sub-) space of availability states and a (sub-) space of unavailability states. The binarization of a multistate system will be introduced with the help of two additional notions: the critical states and the critical transitions. The mathematical modeling of the availability and the reliability of the multistate systems and their binarization gives birth to the notions of the equivalent parameters, that is, the equivalent failure and equivalent repair rates.

In Chapter 3, we treat the modeling of the matrix-like system $(n \times l)$, with $n$ elementary components in-series, which are repeated in-parallel $l$ times. This system is found, quite often, in the electronic devices of detection and signal analysis systems, active and passive redundant systems and manufacturing processes. Quite often too, the $n \times l$ involved components are identical. Analysts may be interested in matrix systems for two different, but not decoupled, reasons. The first concerns the modeling of the availability and the reliability of the systems in the extension of Chapter 2 after a direct adaptation to the matricial structure. The second reason is rather for the modeling of the performance degradations of a system (matrix-like or

1 When it seems appropriate, we will indifferently use the symbols "∨" or "+" to designate the OR-operator and "∧" or "·" to designate the AND-operator throughout the book.

not) if the operation of the system is rather classified in phases defined by thresholds which is determined by the number of the available or the unavailable elementary components. This is exactly the case of distributed systems such as networks which are structured in nodes and arcs. In these types of systems, analysts are more interested in modeling the performance of the network and its evolution from one performance phase to another, more than in the overall availability and reliability of the system. In other words, analysts are interested in transitions from and sojourns in different performance phases. In this approach, the performance states of the distributed system are not simply classified into availability space and unavailability space. The performance states are classified according to the number of available or unavailable elementary components.

Chapters 4 and 5 are devoted to the application of the concepts and notions already introduced above on simple systems in the sense of the structural simplicity of the system and the numerical simplicity of the elementary components' parameters. The structural simplicity results from the small number of elementary components belonging to the system and the simplicity of their logical links. The numerical simplicity results from the following assumption: the elementary components are mutually independent, with time-independent failure and repair rates. This simplicity is purposely required as we assume that the training of the readers of the book should be done through active learning and hand-working on the proposed application cases.

In Chapter 4, we proceed for the systems with redundancy. Redundancy is a technical solution that generally improves system's availability and reliability, but at a higher total cost of installation and maintenance. So, we are talking about systems with $^{n}/_{N}$ redundancy. In the literature, there are two complementary definitions of the system $^{n}/_{N}$. Either it is a macro system made up of $N$ systems that are functionally identical, of which the availability of at least $n$ is sufficient to make the macro system available. Or, it is a macro system made up of $N$ systems that are functionally identical, of which the unavailability of at least $n$ systems is enough to make the macro system unavailable. In the first definition, $n$ indicates the minimum availability condition of the macro system's availability. In the second, $n$ indicates the minimal unavailability condition of the macro system's unavailability. The analyst should define which of the two stand points of view is the most appropriate to his/her analysis.

In Chapter 5, we develop some models for general systems. Assessing the reliability and the availability of general systems is among the most common activities in all the areas relating to systems engineering, such as design and concept validation, maintenance management and optimization, safety management, assessments of periodic testing strategy and operational safety (in normal and accident situations). Dynamic modeling of the availability and reliability of a system goes through several stages: modeling the availability and reliability of the elementary components, the construction of the functional space of the system, the identification of the states associated with the availability space and those associated with the unavailability space, the identification of the critical states, the determination of the equivalent transition rates and, finally, the determination of the availability/unavailability and the reliability/unreliability of the system. A didactic application case is illustrated. It is a simple system consisting of five elementary components.

In Chapter 6, we deal with the modeling of the sequential events. The modeling of sequential events occupies a specific place in the fields of accident analysis and system safety analysis. Often, analysts seek to identify the sequences of failures (systemic, human, procedural, etc.) that may produce serious accidents at the earliest stages of system design. They seek to eliminate or reduce the occurrence of dreaded sequences to interrupt the occurrence order of the events of the sequence and/or to mitigate their consequences. Analysts thus use several analysis tools, including the event tree analysis. Unlike the fault tree, the event tree considers the occurrence order of the events. In this chapter, we have developed two models fairly practiced in the modeling of sequential events: the dynamic model and the static model. Generally, the dynamic model is used more in accident analyses. On the other hand, the static model is widely used in system safety analyses. An academic case composed of four sequential events is treated. A comparative analysis between the models (dynamic and static), illustrated by numerical results, is also carried out.

In Chapter 7, numerical simulation by the Monte Carlo method is briefly presented and illustrated through some applications. We have emphasized on the notion of bijective analytical sampling, the generic approach (the game), the scoring and the statistical processing of the scored figures (standard deviation, variance, uncertainty, etc.) with applications in two simple cases.

The first case describes the evolution over time of a system represented by its graph of states. The transitions between the states are numerically simulated to mainly determine the average sojourn time per state and the number of transitions to each state within a given time interval. The second case concerns a system described by its elementary components and its logical structure. Its reliability and its operating histogram are mainly determined from the numerical simulation of the operating histograms of its elementary components. We have not covered several important topics in the Monte Carlo simulation method to note the non-analytical sampling, the acceleration techniques and the propagation of uncertainties. However, the basic knowledge presented in the chapter allows the readers to independently complete their knowledge of these subjects according to their professional needs in real time.

Finally, mathematical modeling would not be of any engineering interest without acquiring some knowledge related to failure and repair data issued from the real system engineering world. Chapter 8 will therefore be dedicated to the acquisition of these data and their processing via either life-test campaigns or operational experience feedback analyses. The chapter presents the physical testing techniques used for system reliability estimation. These tests make it possible to obtain data relevant to the reliability of the designed or manufactured systems. When a new prototype includes innovations, data are often unavailable or even nonexistent. Subsequently, it is necessary for design optimization purposes to build a knowledge base using physical tests outcome. This makes it possible to estimate the reliability of the components as a function of the operational stresses to which they may be subjected.

Among the different tests used to estimate the reliability of industrial products, we underline the accelerated tests, the aggravated tests and the Bayesian tests.

# 1

# Basic Notions

## 1.1. Introduction

We introduce in this chapter some basic notions that engineers use to perceive the operation of a system. Understanding these basic notions allows us to build mathematical models as close as possible to the real operation patterns of the system. These notions are of two different natures: logical and probabilistic.

Sections 1.2 and 1.3 briefly present these basic notions that the reader must acquire to be able to continue his/her navigation in the book. The basic notions are presented without demonstrations.

In section 1.4, we present the "functional" quality as being the only quality that the system can endorse, disregarding other measurable qualities of the system. Only functional qualities will intervene in the definition and, subsequently, in the modeling of the system operational patterns. Considering the system as a "functional entity" highlights the binary nature of the system operational patterns. This is the primary quality of the system. This functional binarity will be perfectly characterized by measurable quantities issued from the real world of systems as perceived by the engineer. These are precisely the system's failure and repair rates.

In section 1.5, we develop mathematical models describing the evolution over time of the functional performance of the system because of a binary functional concept. These models give us the possibility of characterizing the operation of the system by a multitude of measurable secondary qualities. Each of these secondary qualities will be detailed in the following sections.

Section 1.6 will therefore be devoted to a secondary quality called "availability", being an instantaneous quality by definition. It will be described by a first-order linear differential equation, generally inhomogeneous with variable coefficients.

Section 1.7 will be devoted to other secondary qualities called "reliability" and "maintainability". Being integral qualities, they are described by homogeneous first-order linear differential equations, generally with variable coefficients.

Section 1.8 will be devoted to the modeling of the performance of a binary system. The successive transitions between the state of availability and the state of unavailability will be described as a stochastic process.

Section 1.9 will be devoted to a brief introduction of the multistate system. The simplest multistate system that one can conceive is obviously composed of two binary systems.

## 1.2. Logical notions

Let $\mathcal{L}_n$ be a set of components $\{C_1, C_2, C_3, \dots, C_n\}$ whose states can be described by a set of binary variables $\{X_1, X_2, X_3, \dots, X_n\}$ such as:

$$X_i = \begin{cases} 0, & \text{if the state of component } C_i \text{ satisfies certain conditions} \\ 1 & \text{otherwise.} \end{cases}$$

Let $Y_n$ be the state binary function of the set $\mathcal{L}_n$ such as:

$$Y_n = \begin{cases} 0, & \text{if the state of } \mathcal{L}_n \text{ satisfies certain conditions} \\ 1 & \text{otherwise.} \end{cases}$$

For example, let $\mathcal{L}_4$ be a set of four components $\{C_1, C_2, C_3, C_4\}$. If the state of a component $C_i$ satisfies certain required conditions, its state variable $X_i$ takes the value 0, otherwise 1. Besides, if the binary variables of at least three components are 0, the binary function $Y_4$ takes the value 0, otherwise 1, see Table 1.1.

The binary function $Y_4$ can also be described by the following Boolean expression:

$$Y_4 = X_1 \cdot X_2 \cdot X_3 + X_2 \cdot X_3 \cdot X_4 + X_3 \cdot X_4 \cdot X_1 + X_4 \cdot X_1 \cdot X_2,$$

where the operators "·" and "+" designate, respectively, the Boolean operators "AND → ∧" and "OR → ∨".

REMARK.– Any conjunction, such as $(X_1 \cdot X_2 \cdot X_3)$, can also be called "cut" in system reliability, regardless of the conjunction order (the number of its members).

| | 1 | 2 | 3 | 4 | 5 | 6 | 7 | 8 | 9 | 10 | 11 | 12 | 13 | 14 | 15 | 16 |
|---|---|---|---|---|---|---|---|---|---|---|---|---|---|---|---|---|
| $X_1$ | 0 | 1 | 0 | 1 | 0 | 1 | 0 | 1 | 0 | 1 | 0 | 1 | 0 | 1 | 0 | 1 |
| $X_2$ | 0 | 0 | 1 | 1 | 0 | 0 | 1 | 1 | 0 | 0 | 1 | 1 | 0 | 0 | 1 | 1 |
| $X_3$ | 0 | 0 | 0 | 0 | 1 | 1 | 1 | 1 | 0 | 0 | 0 | 0 | 1 | 1 | 1 | 1 |
| $X_4$ | 0 | 0 | 0 | 0 | 0 | 0 | 0 | 0 | 1 | 1 | 1 | 1 | 1 | 1 | 1 | 1 |
| $Y_4$ | 0 | 0 | 0 | 1 | 0 | 1 | 1 | 1 | 0 | 1 | 1 | 1 | 1 | 1 | 1 | 1 |

**Table 1.1.** *The values of the variables $X_i$ and the function $Y_n$*

In the Boolean expression above, the function $Y_4$ is expressed by four cuts that satisfy the condition defining its binary values for all possible cuts. These are called the "minimal cuts".

We may note that minimal cuts are mutually dependent, whatever the Boolean expression of the function $Y_4$ is.

Boolean algebra is a widely used tool in probabilistic calculations. We will then recall the fundamental laws and axioms of Boolean algebra in the following.

### 1.2.1. *Axioms*

Let $\{X_i, X_j, X_k\}$ be a set of three Boolean variables, $\{\bar{X}_i, \bar{X}_j, \bar{X}_k\}$ be the set of their negations, and the conjectures $\emptyset$ and I be, respectively, the empty and the universal sets. Boolean algebra operates on these variables by two operators (AND) and (OR) denoted by "·" and "+", respectively. We note the following axioms:

1) commutativity:

$$X_i \cdot X_j = X_j \cdot X_i$$

$$X_i + X_j = X_j + X_i.$$

2) Distributivity:

$$X_i \cdot (X_j + X_k) = (X_i \cdot X_j) + (X_i \cdot X_k)$$

$$X_i + (X_j \cdot X_k) = (X_i + X_j) \cdot (X_i + X_k).$$

3) Associativity:

$$X_i \cdot (X_j \cdot X_k) = (X_i \cdot X_j) \cdot X_k$$

$$X_i + (X_j + X_k) = (X_i + X_j) + X_k.$$

4) Idempotence:

$$X_i \cdot X_i = X_i$$

$$X_i + X_i = X_i$$

5) Complementation:

$$X_i + \overline{X_i} = \text{I}.$$

6) Absence of contradiction:

$$X_i \cdot \overline{X_i} = \emptyset.$$

The previous list contains the most basic axioms of the Boolean algebra. They will be recalled throughout the book. Some elementary law can then be derived from the axioms cited above. They are presented in the next section.

### 1.2.2. *Elementary laws*

1) $\overline{\overline{X_i}} = X_i$;

2) $\overline{(X_i + X_j)} = \overline{X_i} \cdot \overline{X_j}$,

3) $\overline{(X_i \cdot X_j)} = \overline{X_i} + \overline{X_j}$,

4) $X_i \cdot (X_i + X_j) = X_i$,

5) $X_i + (X_i \cdot X_j) = X_i$,

6) $X_i \cdot (\overline{X_i} + X_j) = X_i \cdot X_j$,

7) $X_i + (\overline{X_i} \cdot X_j) = X_i + X_j$,

8) $X_i + X_j + X_k = X_i + (\overline{X_i} \cdot X_j) + (\overline{X_i} \cdot \overline{X_j} \cdot X_k)$.

## 1.3. Probabilistic notions

In system reliability, we are interested in the occurrence of random events that can degrade the operation of systems, or even cause them to fail. The occurrence of random events can be described in two ways: either by the probability of occurrence $D(t)$ or by the probability density $\rho(t)$ of occurrence of the concerned undesirable events. Any other probabilistic quantity used in system reliability is a function of these basic quantities.

We will therefore limit this reminder to the probabilistic notions mostly used in system reliability and more specifically in this book.

### 1.3.1. *Probability $D(t)$ and probability density $\rho(t)$*

Let us consider the failure probability $D(t)$ of a system, defined as "the probability of a system to fail within the interval of time $[0, t]$". If the failure probability $D(t)$ exists and is continuous over time, the failure probability density $\rho(t)$ is determined by:

$$\rho(t) = \frac{d}{dt} D(t).$$

That is also equivalent to:

$$D(t) - D(t_0) = \int_{t_0}^{t} \rho(\xi) d\xi .$$

We may also be interested in the survival probability $R(t)$ of the system defined as "the probability of the system not to fail within the interval of time $[0, t]$". it is obvious that it can be described as follows:

$$R(t) = 1 - D(t).$$

Thus, the probability density as a function of the survival probability can be described also by:

$$\rho(t) = -\frac{d}{dt} R(t).$$

### 1.3.2. *Failure occurrence rate* $\lambda(t)$

We can also be interested in determining the occurrence rate $\lambda(t)$ of failures within an infinitesimal interval $\Delta t$ in the vicinity of the instant $t$. The rate of failure occurrence in the vicinity of $t$ is therefore determined by:

$$\lambda(t) = \lim_{\Delta t \to 0} \left[ \frac{1}{1-D(t)} \times \frac{D(t+\Delta t)-D(t)}{\Delta t} \right] = -\frac{1}{R(t)} \times \frac{dR(t)}{dt} = \frac{\rho(t)}{R(t)} .$$

We can therefore describe the survival probability $R(t)$ as a function of the failure rate, given that:

$$\lambda(t) = -\frac{1}{R(t)} \times \frac{dR(t)}{dt}.$$

We can then conclude that:

$$R(t) = \mathrm{Exp}\left(- \int_0^t \lambda(\xi) d\xi\right).$$

We can also describe the survival probability density $\rho(t)$ as a function of the failure rate, such as:

$$\rho(t) = \lambda(t). R(t) = \lambda(t). \mathrm{Exp}\left(- \int_0^t \lambda(\xi) d\xi\right).$$

From now and then, the survival $R(t)$ probability and the non-survival probability $D(t)$ will be referred to as the reliability and the unreliability, respectively, in the following chapters.

### 1.3.3. *Independent random events*

Let $\{A, B, C\}$ be a set of independent random events where each event is defined by its probability of occurrence $\{p(A), p(B), p(C)\}$, then:

1) The conjunction of independent events; the probability $p(A \cdot B)$ that the two random and independent events $(A)$ and $(B)$ occur is determined by:

$$p(A \cdot B) = p(A).p(B),$$

where the conjunction $(A \cdot B)$ is true if and only if the events $(A)$ and $(B)$ are true.

2) The disjunction of independent events; the probability $p(A + B)$ that at least one of the two random and independent events $(A)$ and $(B)$ takes place is determined by:

$$p(A + B) = p(A) + p(\bar{A}).p(B),$$

where the disjunction $(A + B)$ is true if at least one of the events $(A)$ or $(B)$ is true.

3) The disjunction of conjunctions of independent events; or the assertion $\mathcal{L}$ defined by the following Boolean expression:

$$\mathcal{L} = (A \cdot B) + (B \cdot C) + (C \cdot A).$$

The assertion $\mathcal{L}$ can also be rewritten in an equivalent form $\mathcal{L}^*$ such as:

$$\mathcal{L}^* = (A \cdot B) + \left(\overline{(A \cdot B)} \cdot B \cdot C\right) + \left(\overline{(A \cdot B)} \cdot \overline{(B \cdot C)} \cdot C \cdot A\right).$$

After the logical treatment of each conjunction belonging to the equivalent assertion $\mathcal{L}^*$, it comes to be defined by:

$$\mathcal{L}^* = (A \cdot B) + (\bar{A} \cdot B \cdot C) + (\bar{B} \cdot C \cdot A).$$

The assertion $\mathcal{L}$ has six possible equivalent expressions $\mathcal{L}^*$, $(3! = 6)$.

The assertion $\mathcal{L}$ contains minimal but dependent conjunctions. The equivalent assertion contains $\mathcal{L}^*$ disjoint conjunctions. Then, the probability $p(\mathcal{L})$ that the assertion $\mathcal{L}$ is true can be determined by:

$$p(\mathcal{L}) = p(A).p(B) + p(\bar{A}).p(B).p(C) + p(\bar{B}).p(A).p(C).$$

The use of the disjoint conjunctions is interesting in many probability calculations. This property will be often recalled in the following chapters.

## 1.4. System functional description

The general pattern of system operation is defined by successive periods representing system availability and unavailability periods. The observed durations of the availability and unavailability periods are random. This first observation naturally suggests that the models describing the functioning of the systems can only be probabilistic.

Consider a simple electrical switch. The function of the switch is "to establish electrical contact and interrupt it on demand". The electric cables connected inside the switch may be well fixed or badly fixed. However, the switch continues to fulfill its function of "establishing electrical contact and interrupting it on demand". Compared to a conceptually perfect state, the switch is in a degraded state, but it continues to perform its function. The switch can become out of service by the loss of physical contact with its cables or decision of the operator in compliance with the safety rules to prevent potential accidents. A good operator waits neither until the occurrence of the physical damage of the system nor until the functional degradation of the system becomes a source of safety or security concern.

A system can therefore continue to be available to carry out its function, even in degraded modes, up to a certain limit of its degradation before it becomes functionally unavailable. Or, it is switched to the unavailability state upon a decision by the operator before the occurrence of the systemic failure. The critical degradation limits are quite often imposed for safety or security reasons.

What do we see when observing the functioning of the systems around us? Leaving aside specific qualities of the system, such as its conceptual or functional complexities, interactivity or proactivity, modernity or the archaism of its design. The most basic qualities to note when observing the functioning of a system are as follows:

– its ability to be available. This ability is called “availability”;

– its ability to remain available for a given time period without failure. This ability is called “reliability”.

The basic performance of a system can therefore be described by the repetition of a cycle defined by two intervals:

– an operating interval ($\tau_f$), before the system breaks down;

– a non-operating interval ($\tau_r$), before the system is back to operation.

The operational experiences feedback demonstrates that the durations of these successive intervals ($\tau_f$, $\tau_r$) are random.

The analysts represent this availability/unavailability alternation by means of an operation histogram. The operation histogram is a graphical tool of great interest. We present in Figure 1.1 a typical operation histogram in the simplest form, that is, in a binary form, available and unavailable. The binary operation histogram allows us to average the availability and the unavailability of a system over a given time interval. But it does not allow us to perform a more detailed performance analysis. For example, if the system is available, is it in a perfect or a degraded working order?

This is natural because “being available” only means being available to carry out a mission/function. This, in no way, indicates whether the system is functioning perfectly or in degraded mode. The concepts “availability” and “reliability” are two functional concepts resulting from a functional analysis of the performance of a system. The two qualities (availability and reliability) are of a binary nature: “available or not” and “reliable or not”. Consequently, the “system” concept itself is also a “functional” concept before being “physical”. In reality, a system is made up of physical, functional, procedural, and regulatory entities, all connected in a logical structure to achieve a well-defined mission.

These binary natures are shared by all systems, regardless of their concept, complexity or mission. They are of a binary and probabilistic nature. We are going to study these aptitudes of the systems to extract the basic properties that are shared by all systems.

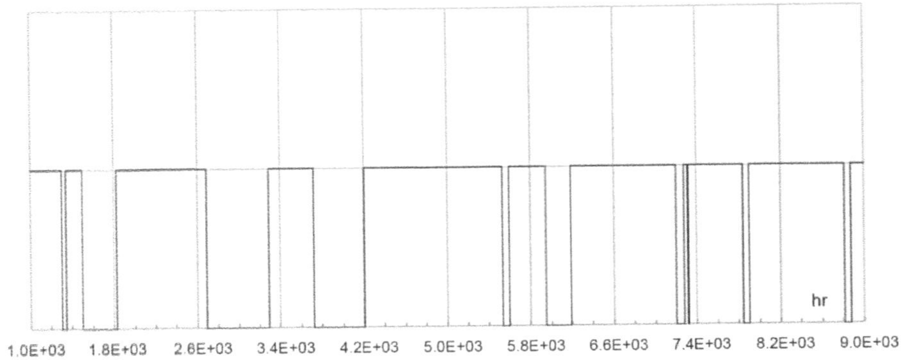


**Figure 1.1.** *Representative binary operation histogram*

## 1.5. Binary system and basic model

Let $S$ be a binary system, whose functional space contains only two subspaces of a single state each; the availability subspace, which contains the state (0), and the unavailability subspace, which contains the state (1) (Figure 1.2). The binary system is a basic concept where only two transitions are possible: failure and repair/recovery.

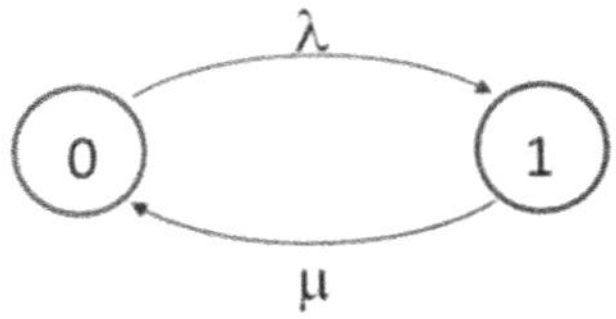


**Figure 1.2.** *The binary system and its states. For a color version of this figure, see www.iste.co.uk/elhami/reliability.zip*

The failure of a binary system is described by a transition from the state of availability (0) to the unavailability state (1). This is true regardless of the quality of the "operating" state or the severity of the "unavailability" state.

This is always true as the notions of "functional degradation" and "severity of failures" do not exist in a binary systems by definition. The "failure rate" $\lambda$ characterizes the transition toward the unavailability state (1), and the "repair rate" $\mu$ characterizes the transition toward the availability state (0). The designation of 0 and 1 for availability and unavailability, respectively, is conventional. The analyst is free to choose the adequate convention.

The transition rates $(\lambda, \mu)$ may be constant or any function of time: continuous or discontinuous. This aspect will be dealt with in detail later. These transition rates constitute the basic properties of all systems, in general: binary and multistate ones. The basic properties being identified and defined, we must then deal with the primary qualities of the systems: the availability in [1.6] and the reliability in [1.7].

A third characteristic property is still necessary to complete the characterization of systems. We will also talk about this later.

## 1.6. The availability of binary systems

"Availability" is the ability of the system, at a given instant $t$, to carry out a given mission under well-defined conditions[1]. It is measured by the probability $p(t)$ of being available at instant $t$.

From this definition, we will develop a model that describes the availability of a binary system. Let $p(t)$ be the availability at the time $t$ of a binary system, and $p(t+\Delta)$ is the availability of the system at time $t+\Delta$ in the vicinity of $t$. The availability $p(t+\Delta)$ can be determined as follows:

$$p(t+\Delta) = [1-\lambda(t)\Delta]\cdot p(t) + \mu(t)\Delta\cdot[1-p(t)],$$

where $(1-\lambda(t)\Delta)\cdot p(t)$ is the probability that the system is available at the time $t$ and does not fail during the interval $\Delta$; $\mu(t)\Delta\cdot(1-p(t))$ is the probability that the system is unavailable at time $t$ and that it recovers during the interval $\Delta$; $\Delta$ is a time interval that is as small as possible, that is, $\Delta \to 0$.

1 The international standard ISO/IEC 2382-14.

If $p(t)$ is a continuous function in the vicinity of $t$, the above equation can also be rewritten as follows:

$$p(t) + \frac{dp(t)}{dt}\Delta + \left[\frac{d^2p(t)}{dt^2}\frac{\Delta^2}{2!} + \cdots\right] \\ = p(t) - [\lambda(t) + \mu(t)]\Delta p(t) + \mu(t)\Delta.$$

Finally, we find the differential equation that describes the evolution of the availability over time as $\Delta \to 0$:

$$\frac{d}{dt}p(t) = -[\lambda(t) + \mu(t)]p(t) + \mu(t), \qquad [1.1]$$

where $\lambda(t)$ and $\mu(t)$ are, respectively, the failure rate and the repair rate of the system. They can be constant or any function of time. The only assumption made to obtain equation [1.1] is that availability and its first derivative are continuous in the vicinity of time $t$.

Equation [1.1] is a first-order linear differential equation, generally inhomogeneous ($\mu(t) \neq 0$). Its solution requires the knowledge of an initial condition, that is, the initial availability $p_0$. In practice, $p_0$ is always expressed as a function of the initial unavailability $\gamma$ as follows:

$$p_0 = 1 - \gamma,$$

where the initial unavailability, $\gamma$, is often called the "refusal to start probability".

Note that equation [1.1] always evolves toward an asymptotic value. The asymptotic value $p_\infty$ of the availability is determined as follows:

$$p_\infty = \frac{\mu_{t\to\infty}(t)}{\mu_{t\to\infty}(t)+\lambda_{t\to\infty}(t)}.$$

Equation [1.1] describes the evolution over time of any system, whatever its complexity. This includes binary and multistate systems as long as the system is perfectly defined by its characteristic parameters $(\lambda, \mu, \gamma)$. It is necessary to emphasize that $\lambda$ and $\mu$ are transition rates; however, $\gamma$ is a probability. All the parameters have positive values. The parameter $\gamma$ is always less than or equal to 1, while $\lambda$ and $\mu$ are unlimited.

Note also that the asymptotic value of the availability $p_\infty$ of a non-repairable system ($\mu = 0$) is always zero.

The general solution of equation [1.1] can be put in the form:

$$p(t) = e^{-\int_0^t (\lambda(\xi)+\mu(\xi))d\xi} \times \left[\left(\int_0^t \mu(\xi). e^{\int_0^\xi (\lambda(\eta)+\mu(\eta))d\eta} d\xi\right) + p_0\right]. \qquad [1.2]$$

SPECIAL CASE.– If, and only if, the failure and repair rates $(\lambda, \mu)$ are constant over time, the solution of equation [1.2] can take the following form:

$$p(t) = \frac{\mu}{\lambda+\mu}\left(1 - e^{-(\lambda+\mu)t}\right) + p_0 e^{-(\lambda+\mu)t}. \qquad [1.3]$$

CONVENTION.– From this section and thereafter, we will use the term "component" only to describe any binary functional entity whose characteristic parameters $(\lambda, \mu, \gamma)$ are well defined whether $(\lambda, \mu)$ are time constants or time variables.

Subsequently, the term "system" will be reserved exclusively for multistate functional entities whose components are well identified and whose characteristic parameters $(\lambda_{sys}, \mu_{sys}, \gamma_{sys})$ should be determined.

The "component" or the "system" status is perfectly conventional. It is decided by the analyst, depending on the granularity of his/her functional analysis and the availability of the basic data $(\lambda, \mu, \gamma)$.

Finally, if we should develop a model describing the unavailability rather than availability, we can follow exactly the same approach as for the availability above. Let $q(t)$ be the unavailability at time $t$ of a binary system, then $q(t + \Delta)$ is the unavailability of the system at time $t + \Delta$ in the vicinity of $t$, and it will be determined as follows:

$$q(t + \Delta) = [1 - \mu(t)\Delta] \cdot q(t) + \lambda(t)\Delta \cdot [1 - q(t)],$$

where $(1 - \mu(t)\Delta) \cdot q(t)$ is the probability that the system is unavailable at time $t$ and that it remains so during the interval $\Delta$, $\lambda(t)\Delta \cdot (1 - q(t))$ is the probability that the system is available at time $t$ and fails during the interval $\Delta$ and $\Delta$ is a time interval that is as small as possible, that is, $\Delta \to 0$.

By taking the limit when $\Delta \to 0$, we find the following expression which describes the unavailability in the time:

$$\frac{d}{dt}q(t) = -[\lambda(t) + \mu(t)]q(t) + \lambda(t). \qquad [1.4]$$

Note that equation [1.4] always evolves toward an asymptotic value. The asymptotic value $q_\infty$ of the unavailability can be determined from equation [1.4] as follows:

$$q_\infty = \frac{\lambda_{t\to\infty}(t)}{\mu_{t\to\infty}(t)+\lambda_{t\to\infty}(t)}.$$

Note also that the asymptotic value of the unavailability $q_\infty$ of a non-repairable system ($\mu = 0$) is always equal to 1.

EXAMPLE 1.1.–

Prove that equation [1.5] correctly describes the unavailability $q(t)$ of a well-defined component whose transition rates $(\lambda, \mu)$ are constant. Use the proposed approach to develop the availability model in equation [1.1]:

$$q(t) = \frac{\lambda}{\lambda+\mu}\left(1 - e^{-(\lambda+\mu)t}\right) + \gamma e^{-(\lambda+\mu)t}. \qquad [1.5]$$

Let C1 and C2 be two components whose parameters ($\lambda$, $\mu$, $\gamma$) are well defined, such as: C1 $(10^{-4}\text{h}^{-1}, 1\,\text{h}^{-1}, 3\times10^{-1})$ and C2 $(10^{-2}\text{h}^{-1}, 2\times10^{-2}\text{h}^{-1}, 0)$. We trace the evolution over time of their respective unavailability in Figure 1.3.

Note that if the two components C1 and C2 can perform the same function, we can design a system $S$ composed of these two components. System $S$ could start up functioning with component C2, at $t = 0$ h. Then, it switches from the component C2 to the component C1 at instant $t = 3$ h, using an adequate switch device. The system $S$ would perform the same function but with better availability at the start-up and in the medium and long time intervals.

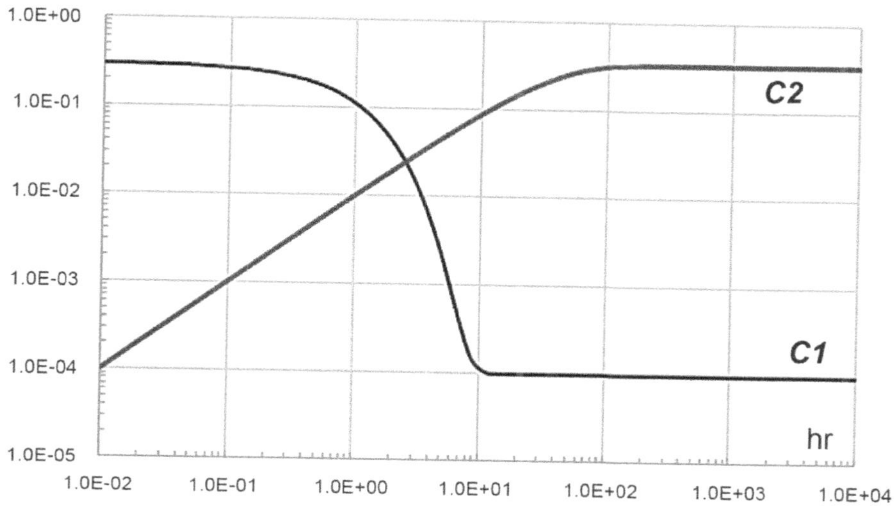


**Figure 1.3.** *The evolution over time of the unavailability of components C1 and C2. For a color version of this figure, see www.iste.co.uk/elhami/reliability.zip*

## 1.7. The sojourn probabilities

We devote this section to the development of models describing the probabilities of sojourn in the availability and the unavailability states.

The term "sojourn probability" is general, as it can target the sojourn in a specific state (availability, unavailability, or degraded in case of multistate systems). Indeed, the term "sojourn probability" has meaning for multistate systems only. In this chapter, which deals exclusively with binary systems, there are only two states: availability and unavailability.

So, when it comes to sojourns in availability or unavailability states of a binary system, analysts use the terms: reliability and maintainability.

### 1.7.1. *Reliability model*

Reliability is the ability of the system to be available within an interval of time $t$ under well-defined conditions and without breaking down[2]. It is

2 The international standard ISO/IEC 2382-14.

measured by the probability of being available without failure during the interval $t$.

Let $R(t)$ be the reliability of a system during an interval of time $t$ and $R(t+\Delta)$ is reliability during the interval $t+\Delta$. We can therefore describe $R(t+\Delta)$ in terms of $R(t)$ as follows:

$$R(t+\Delta) = (1-\lambda(t)\Delta).R(t),$$

where $R(t)$ is the probability of being reliable during the interval $[0,t]$, $(1-\lambda(t)\Delta)$ is the probability of not failing during the interval $\Delta$ in the vicinity of $t$ and $\Delta$ is a time interval that is as small as possible, $\Delta \to 0$.

Given that $R(t)$ is a continuous function over time, the above equation can also be put in the following form:

$$R(t) + \frac{dR(t)}{dt}\Delta + \left[\frac{d^2R(t)}{dt^2}\frac{\Delta^2}{2!} + \cdots\right] = R(t) - \lambda(t)\Delta.R(t).$$

Thus, we find the differential equation that describes the evolution of reliability over time:

$$\frac{d}{dt}R(t) = -\lambda(t)R(t), \qquad [1.6]$$

where $\lambda(t)$ is the failure rate. It can be constant or any function of time.

Note that if the system is not repairable ($\mu \to 0$), equation [1.1], describing the availability, tends toward equation [1.6] that describes the reliability. In this case, availability and reliability would be described by exactly the same differential equations. However, the variable $t$ would represent a time interval for the reliability, while it would refer to an instant of time for the availability.

Equation [1.6] is a homogeneous first-order differential equation. The reliability of any system during a zero-time interval, $f_0$, is naturally equal to 1, as long as the failure rate has a determined value.

The general solution of equation [1.6] is as follows:

$$R(t) = e^{-\int_{\xi=0}^{t}\lambda(\xi)d\xi}. \qquad [1.7]$$

Reliability does also have an asymptotic value $R_\infty$. The asymptotic value of reliability $R_\infty$ tends to zero, as long as the failure rate has a determined non-zero value:

$$R_\infty = \lim_{t\to\infty}\left(R(t)\right) = -\lim_{t\to\infty}\left(e^{-\int_{\xi=0}^{t}\lambda(\xi)d\xi}\right). \qquad [1.8]$$

In some cases, the reliability function $R(t)$ is known but not the failure rate. Then, we can calculate the failure rate using the following model:

$$\lambda(t) = -\frac{1}{R(t)}\frac{d}{dt}R(t). \qquad [1.9]$$

The reliability function $R(t)$ can also be defined either analytically or numerically in tables.

The most cited models in the literature for failure rates, the corresponding reliability function and the mean time to failure (MTTF) are described in Table 1.2.

| | Failure rate model | Failure rate | Reliability | MTTF |
|---|---|---|---|---|
| 1 | Constant | $\lambda$ | $e^{-\lambda t}$ | $\frac{1}{\lambda}$ |
| 2 | Linear | $\alpha + \beta t$ | $e^{-(\alpha t+\frac{\beta}{2}t^2)}$ | $\frac{2\sqrt{\pi}}{\beta}e^{-\frac{\alpha^2}{2\beta}}$ |
| 3 | Weibull | $\frac{\beta}{\eta}\left(\frac{t}{\eta}\right)^{\beta-1}$ | $e^{-\left(\frac{t}{\eta}\right)^{\beta}}$ | $\eta\Gamma\left(1+\frac{1}{\beta}\right)$ |

**Table 1.2.** *The most cited failure rates, reliability and MTTF*

### 1.7.2. *Maintainability*

If we can conceive that the reliability is the sojourn probability in the state of availability during a time interval equal to $t$, we can by analogy conceive that the sojourn probability in the state of unavailability during a time interval is equal to $t$. This probability is a measure of the system's maintainability.

Let $m(t)$ be the maintainability of a system during an interval of time $t$ and $m(t+\Delta)$ its maintainability during the interval $t+\Delta$, in the vicinity of instant $t$. One can therefore describe $m(t+\Delta)$ in terms of $m(t)$ as follows:

$$m(t+\Delta) = (1-\mu(t)\Delta).m(t),$$

where $m(t)$ is the probability of being in the unavailable state during the interval $[0,t]$, $(1-\mu(t)\Delta)$ is the probability of remaining unavailable during the interval $\Delta$ and $\Delta$ is a time interval that is as small as possible, that is, $\Delta \to 0$.

As $m(t)$ is a continuous function over time, the above equation can also be put in the following form:

$$m(t) + \frac{dm(t)}{dt}\Delta + \left[\frac{d^2m(t)}{dt^2}\frac{\Delta^2}{2!} + \cdots\right] = m(t) - \mu(t)\Delta.m(t).$$

Thus, we find the differential equation that describes the evolution of maintainability over time:

$$\frac{d}{dt}m(t) = -\mu(t)m(t), \qquad [1.10]$$

where $\mu(t)$ is the repair rate. It can be constant or any function of time.

It is important to note the confusion between the "unavailability" state and the "maintenance" state. This confusion results quite naturally from the nature of the "functional analysis" itself. This confusion would be removed if we carry out a "dysfunctional analysis" instead. Note that the dysfunctional analysis approach would create the same confusion between being "available" and "functioning".

As far as our subject is the "functional analysis", we are underling that "unavailability" is not always related to "reparation or maintenance", in the strict sense of reparation and maintenance. Subsequently, what we call the repair rate is not in fact an inherent reparation rate of the system. It is in most cases a recovery rate, in the sense of "recovery" from the unavailability state.

Note also the non-proportionality between the huge efforts devoted to developing failure rate models and databases for reliability purposes and the few efforts devoted to developing repair rate models and databases for maintainability by reliability engineering community.

The most cited models for the repair rate are mainly the constant repair rate and, far behind, the constant repair time.

## 1.8. Failure, repair and stochastic processes

Practices show us that the duration of sojourns in the "availability" state and the duration of sojourns in the "unavailability" state of a system are random variables. These variables can therefore be described because of the use of certain stochastic processes.

Among the most widely implemented processes in system reliability modeling, simulation, and analysis (MSA), we can cite for example the Poisson stochastic process. The Poisson stochastic process (PSP) is characterized by time-constant transition rates. We can identify two classes of processes: renewal and non-renewal.

The occurrence of smoke detector false alarms is a renewal process, as detector initialization is almost instantaneous after each false alarm. If the occurrence rate of the false alarms is constant over time, we can say that the false alarms follow a stochastic renewable Poisson process.

One cannot say the same regarding the true alarm signal. The true alarm is restrictively correlated to the smoke occurrence. It follows the stochastic occurrence pattern of the smoke occurrence if it has any.

The occurrence of the "failure to detect smoke" by the detector is a non-renewable process since the reparation/maintenance of the detector is not instantaneous after each smoke detection failure. For some types of detectors, it even takes a relatively long repair/replacement time before the detection system can go into service again. This is exactly the case with repairable systems. A repairable system cannot fail if it is already in reparation, and it cannot be in reparation if it has not yet failed.

But the cycle "availability-unavailability" itself is a renewable stochastic process for all reparable systems.

### 1.8.1. *Renewable stochastic processes*

By "renewable stochastic process", we mean the stochastic realizations which can be repeated, without conditions. Consider $D$ a smoke detector whose occurrence rate of false alarms $\omega$ is constant over time. Describe the probability $Q_n(t)$ of observing $n$ false alarms, in the interval $[0, t]$, given that detector initialization is instantaneous.

The probability $Q_n(t)$ of observing $n$ "false alarms" in the interval $[0, t]$ can be described by:

$$Q_n(t) = \int_0^t [Q_{n-1}(\xi)](\omega d\xi)\left(e^{-\omega(t-\xi)}\right), \qquad [1.11a]$$

where $Q_{n-1}(\xi)$ is the probability of observing $(n-1)$ "false alarms" in the interval $[0, \xi]$, $\omega d\xi$ is the probability of having a false alarm around the instance $\xi$ and $e^{-\omega(t-\xi)}$ is the probability of having no false alarms in the interval $[\xi, t]$.

Equation [1.11a] can be rearranged as follows:

$$\left(Q_n(t)e^{+\omega t}\right) = \int_0^t \left(Q_{n-1}(\xi)e^{+\omega\xi}\right)(\omega d\xi). \qquad [1.11b]$$

Equation [1.11b] can equally be rewritten as:

$$F_n(t) = \int_0^t F_{n-1}(\xi)\omega d\xi \qquad [1.12]$$

where

$$F_n(t) = Q_n(t)e^{+\omega t}, \quad Q_0(t) = e^{-\omega t}$$

and $F_0(t) = Q_0(t)e^{+\omega t} = e^{-\omega t}e^{+\omega t} = 1$.

By deriving equation [1.12] once, we find that:

$$\frac{d}{dt}F_n(t) = \omega F_{n-1}(t). \qquad [1.13]$$

By differentiating equation [1.13] $n$ times, we find that:

$$\frac{d^n}{dt^n}F_n(t) = \omega^n F_0(t) = \omega^n. \qquad [1.14]$$

Given equation [1.14], we obtain the following solution:

$$F_n(t) = \frac{(\omega t)^n}{n!} = Q_n(t)e^{+\omega t}. \qquad [1.15]$$

Then, the probability of observing $n$ "false alarms", $Q_n(t)$, in the interval $[0, t]$ is determined by:

$$Q_n(t) = \frac{(\omega t)^n}{n!}\ e^{-\omega t}. \qquad [1.16]$$

Equation [1.16] describes well the renewal probability of the same event $n$ times on the same observed individual detector, during an interval $t$, if no renewal condition is required.

The variation of the probability of occurrence $n$ times within a given interval as a function of the duration of the interval is plotted in Figure 1.4.

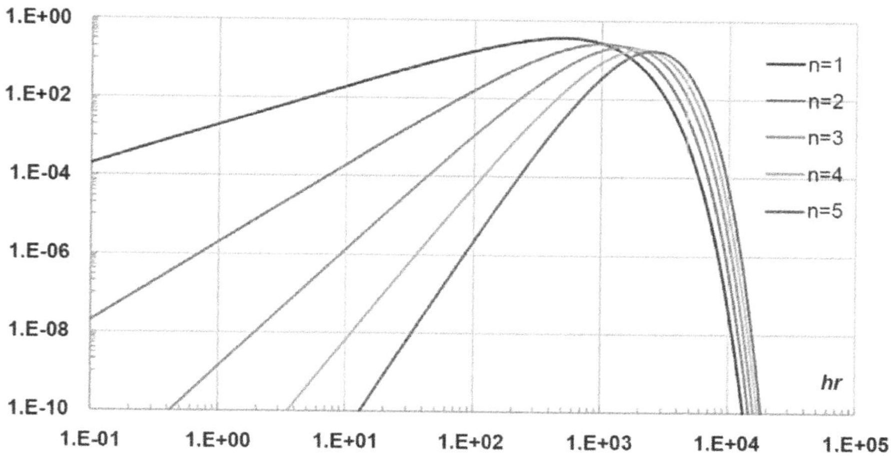


**Figure 1.4.** *The profile of the probability of occurrence $n$ times in a given time interval of a renewal Poisson stochastic process ($\omega = 2 \times 10^{-3}\, h^{-1}$). For a color version of this figure, see www.iste.co.uk/elhami/reliability.zip*

### 1.8.2. *Conditionally renewable stochastic processes*

By "conditionally renewable stochastic process", we mean the stochastic realizations which result, conditionally, from the interference with other well-defined stochastic realizations.

In the following case, it is a single system. The system cannot fail without first being repaired. Nor can it be repaired unless it fails first. However, the failure-repair cycle, itself, is stochastic and renewal.

Thus, instead of observing instantaneously reparable systems, we are interested in determining the probability $P_n(t)$ of having $n$ cycles (failure-repair) within a given time interval $[0,t]$. In the case where the failure and repair rates are constant over time, we can express $P_n(t)$ by:

$$P_n(t) = \int_{\xi=0}^{t} \int_{\eta=\xi}^{t} [\, P_{n-1}(\xi)](\lambda d\xi) e^{-\mu(\eta-\xi)} (\mu d\eta) e^{-\lambda(t-\eta)}, \qquad [1.17a]$$

where $P_{n-1}(\xi)$: the probability of having $(n-1)$ cycles (failure-repair) in the interval $[0,\xi]$; $\lambda d\xi$ is the probability of having a failure around the instant $\xi$, $e^{-\mu(\eta-\xi)}$ is the probability of not being repaired in the interval $[\xi,\eta]$, $\mu d\eta$ is the probability of having the repair around the instant $\eta$ and $e^{-\lambda(t-\xi)}$ is the probability of having no other failures in the interval $[\eta,t]$.

Equation [1.17a] has a general analytical solution ($n > 0$). This general analytical solution of the equation has the following form[3]:

$$P_n(t) = \Psi_n(t) e^{-\lambda t} - \Phi_n(t) e^{-\mu t}, \qquad [1.17b]$$

where

$$\Psi_n(t) = \left(\frac{\lambda\mu}{\sigma^2}\right)^n \left[\sum_{j=0}^{n} (-1)^j \cdot C_j^n \cdot \frac{(\sigma t)^{n-j}}{(n-j)!}\right]$$

3 EID, Mohamed, "Modelling sequential events for risk, safety and maintenance assessments". SSARS 2010, June 20–26, 2010, Gdańsk-Sopot, Poland. JPSRA 2010 VOLUME 1, NUMBER 1, p83-88. ISSN: 2084-5316. Available at: http://ssars.am.gdynia.pl/upload/SSARS2010PDF/VOL1/SSARS2010-EID.pdf.

$$\Phi_n(t) = (-1)^n \cdot \left(\frac{\lambda\mu}{\sigma^2}\right)^n \left[\sum_{j=0}^{n} B_j^n \cdot \frac{(\sigma t)^{n-j}}{(n-j)!}\right]$$

and

$$\sigma = (\mu - \lambda),$$

$$C_0^n \cdot= 1, B_0^n \cdot= 0, n \geq 0,$$

$$C_n^n \cdot= B_n^n, B_n^n \cdot= C_{n-1}^n + B_{n-1}^n, n \geq 1,$$

$$C_{j-1}^n \cdot= C_{j-2}^n + C_{j-1}^{n-1}, B_{j-1}^n \cdot= B_{j-2}^n + B_{j-1}^{n-1}, n \geq 1.$$

The evolution of the probability $P_k(t)$ is represented in Figure 1.5, with values $k$ ranging between 1 and 6, for a system whose failure and repair rates are $\lambda = 8^{-3}$/h and $\mu = 1.5^{-1}$/h, respectively.

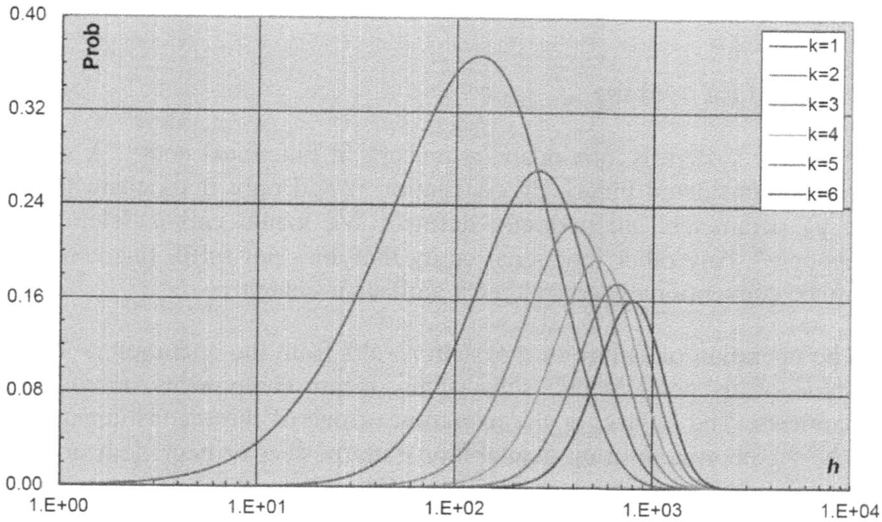


**Figure 1.5.** *The evolution of the probability $P_n(t)$over time, ($\lambda = 8^{-3}/h, \mu = 1.5^{-1}/h$). For a color version of this figure, see www.iste.co.uk/elhami/reliability.zip*

If one is interested in the occurrence of a cycle (failure-repair), equation [1.17a] would take the following form:

$$P_1(t) = \int_{\xi=0}^{t} \int_{\eta=\xi}^{t} \left[e^{-\lambda\xi}\right](\lambda d\xi) e^{-\mu(\eta-\xi)}(\mu d\eta) e^{-\lambda(t-\eta)}. \qquad [1.18]$$

EXAMPLE 1.2.–

Prove that the probability $P_1(t)$ of having a single cycle (failure-repair) in the interval $[0, t]$ is:

$$P_1(t) = \left(\frac{\lambda\mu}{\sigma^2}\right) \cdot \left[\left((\mu - \lambda)t - 1\right)e^{-\lambda t} + e^{-\mu t}\right]. \qquad [1.19]$$

Other stochastic processes are also used by system reliability analysts to describe the occurrence of events with time-dependent occurrence rates. This is the case, for example, for Gaussian processes to propagate uncertainties, Weibull processes that describe the aging of certain systems or gamma processes which are often used to describe the occurrence of certain systems degradation modes.

## 1.9. Multistate system

A binary system is, obviously, a conceptual functional entity. A system would be considered binary by convention if and only if its characteristic $(\lambda, \mu, \gamma)$ parameters are perfectly defined. We would call it "elementary component". Any other functional entity that does not fulfill this condition would be conventionally considered a multistate system.

The operation or failure of a system results from the operation or failure of some of its constituents: subsystems, macro-components, modules or components. The failure, or the operation of one of these constituents, can put the whole system in an availability state (perfect or degraded) or in an unavailability state (partial or complete). In order to be able to assess the reliability and availability of such a system, we must identify its availability and unavailability states. Systems analysts use a set of methods intended for the functional analysis of multistate systems. We briefly outline these methods in Chapter 2.

EXAMPLE 1.3.–

Let $S$ be a system consisting of two independent elementary components $C1$ and $C2$. The system $S$ is then a four-state system (see Figure 1.6):

– the state $E1$ is the perfect availability state, where the two elementary components are available:

$$E1 = C1 \cdot C2;$$

– the state $E2$ and $E3$ are degraded states, where only one elementary component is available:

$$E2 = \overline{C1} \cdot C2 \qquad \text{and} \qquad E3 = C1 \cdot \overline{C2};$$

– the state $E4$ is the state of full unavailability, where the two elementary components are unavailable:

$$E4 = \overline{C1} \cdot \overline{C2}.$$

If it suffices that an elementary component $C1$ or $C2$ makes the system $S$ to fail, this system would experience a single availability state and three unavailability failure states.

If both basic components $C1$ and $C2$ must fail to result the system failure, the system would experience three availability states and a single unavailability state.

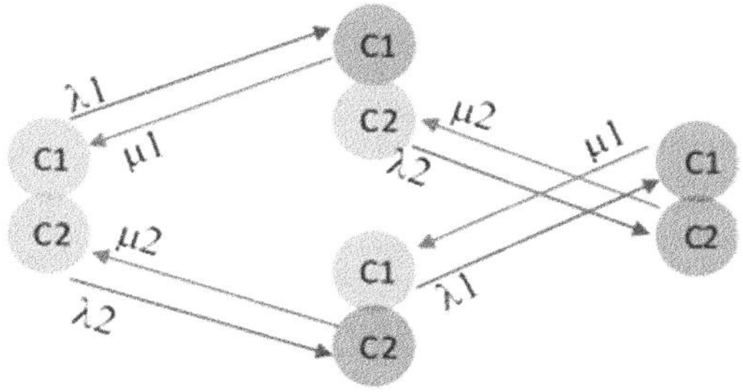


**Figure 1.6.** *A multistate system. For a color version of this figure, see www.iste.co.uk/elhami/reliability.zip*

We will deal with this point in more detail later and introduce some basic notions regarding multistate systems and interstate transitions.

Note that a system consisting of $n$ independent elementary components will contain $2^n$ states shared between a single full availability state, a full unavailability state and $2^{n-1}$ degraded states. In these gray states, the system is either in "degraded availability" or in "partial unavailability". The division of the $2^{n-1}$ gray states between degraded availability and partial unavailability will be done according to the operational criteria that define the mission of the system and its implementation.

# 2

# Modeling of Multistate Systems

The modeling of multistate systems occupies a major place in system reliability and safety studies. Presenting a system by its states, using a state graph, is the method most used by the analysts who develop dynamic models to describe the evolution of the system over time. However, any state graph can also be expressed in the form of a Boolean expression. Thus, we will use state graphs and Boolean expressions indiscriminately to describe the success or the failure of systems.

In this chapter, we rely on the simplest multistate systems that can be designed to introduce some basic notions such as critical states, critical transitions and forbidden transitions. The simplest multistate system may obviously be of two configurations. Either, the system is composed of two elementary components. Subsequently, it is analytically described by four functional states. Or, the system is systemically described by only three functional states.

Thus, we distinguish between two types of multistate systems regardless of their complexity. The first type concerns the system where the transition rates between the states are not raw data but can be determined separately. This is the case of systems whose structures are analytically represented by their elementary components. Subsequently, the fine structure of each state is well known. They will be called "systems with analytical representation (local representation)". The second type concerns systems where only the functional states and the transitions rates between the states are given as raw data, with no knowledge of the states' fine structure. They will be called "systems with systemic representation (global representation)".

In section 2.1, the system with analytical representation is discussed. The notion of a universal functional space consisting of two disjoint subspaces is presented. These two subspaces will be referred to as the "availability space" and the "unavailability space". The terms "success" and "availability" and the terms "failure" and "unavailability" will be interchangeably used in the book.

In section 2.2, we introduce the notion of "critical transitions".

In section 2.3, we exploit critical transitions to introduce the notion of "equivalent transition rates" and then the "binarization" of the system.

In section 2.4, we treat the "failure-to-start" probability of the system to complete its binarization.

In section 2.5, we present "forbidden transitions" as a consequence of the functional independence between the elementary components.

In section 2.6, we briefly introduce the description of functional states using Boolean expressions. We demonstrate the equivalence between state graphs and Boolean expressions.

In section 2.7, we present the multistate system with systemic representation.

In section 2.8, we present the Markov graph model and its solution.

## 2.1. Systems with analytical representation

The simplest possible multistate system will obviously consist of two elementary components[1]. We will reuse the two-component system $S(C_1, C_2)$, already presented previously in example 1.1 (Figure 1.3). The system $S$ has four possible states ($2^2$) independently of the definition of the availability or the associated unavailability. We can then distinguish two possible configurations for the system unavailability, $U1$ and $U2$.

The unavailability mode $U1$ is described as follows: "the unavailability of only one of the elementary components is enough to switch the system's state into unavailable". The unavailability mode $U2$ is described as follows: "the unavailability of the two elementary components is necessary to switch the system's state into unavailable". These two configurations are schematically presented in Figure 2.1.

1 A binary system whose characteristic parameters $(\lambda, \mu, \gamma)$ are well determined.

The universal functional space of the system $S$ always includes four states ($2^2$). It is composed of two subspaces: the "availability" space and the "unavailability" space. The system universal functional space depends neither on the definition of the availability state nor on the definition of the unavailability state of the system. However, the composition of the availability subspace and the unavailability subspace depend on these definitions.

In the case of the unavailability mode $U1$, "the unavailability of only one of the two elementary components is enough to switch the system's state into unavailable", the unavailability subspace of the system comprises three states (Figure 2.1a). However, the unavailability of the type $U2$, "the failure of the two elementary components is necessary to switch the system's state into unavailable", the subspace of the unavailability of the system comprises a single state. In the class of two component systems, we distinguish two classes of systems: the class of systems $S_1$, whose unavailability is described by the unavailability mode $U1$ (Figure 2.1a), and the class of systems $S_2$, whose unavailability is described by the unavailability mode $U2$ (Figure 2.1b). System $S_1$ is often called an in-series system, while $S_2$ is called an in-parallel system in engineering applications.

These definitions are always true for coherent multistate systems. The multistate system is functionally coherent if none of the failures of the elementary components can improve the system performance and if none of the repairs of the elementary components can degrade the system performance.

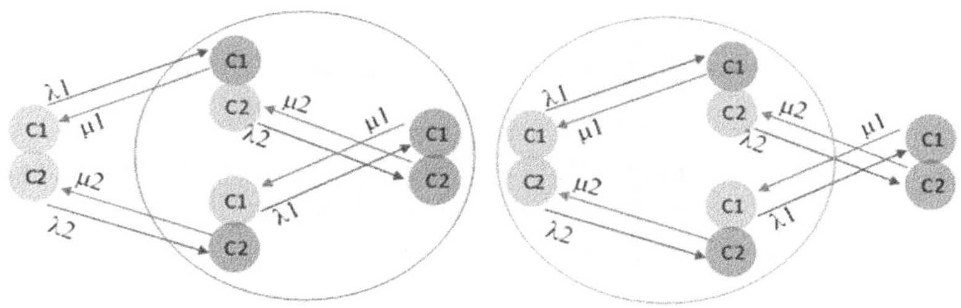


a) b)

**Figure 2.1.** *Two modes of system unavailability (a) System $S_1$ (b) System $S_2$. For a color version of this figure, see www.iste.co.uk/elhami/reliability.zip*

Once the availability and the unavailability subspaces of the system are determined, we can proceed to determine the availability or the unavailability of the system. Notice that we need to determine only one of the two subspaces to estimate both qualities of the system: the availability and the unavailability. The two spaces are exclusively dependent, and the two qualities are to be estimated.

EXAMPLE 2.1.–

Consider a system $S_2$ made up of two elementary components $C1$ and $C2$, which operate in parallel. The system fails if and only if both components fail. Determine and trace the unavailability of the system over time, knowing that the characteristic parameters $(\lambda, \mu, \gamma)$ of the components are:

$$C1(10^{-4}\,h^{-1}, 1\,h^{-1}, 3 \times 10^{-1}) \text{ and } C2(10^{-2}\,h^{-1}, 2 \times 10^{-2}\,h^{-1}, 0).$$

Figure 2.1b is the state graph that represents the functional space of system $S_2$.

The unavailability of $S_2$ is plotted in Figure 2.2. The unavailability of $S_2$ is nothing but the probability of being in the unavailability subspace that includes the state $E_4(\overline{C_1}, \overline{C_2})$ where both $C_1$ and $C_2$ are unavailable. Then, the unavailability $Q_{S_2}$ of $S_2$ is determined by:

$$Q_{S_2}(t) = q_1(t) \cdot q_2(t).$$

While the availability $P_{S_2}$ is the probability of being in the availability space that includes three states: $E_1(C_1, C_2)$, $E_2(\overline{C_1}, C_2)$ and $E_3(C_1, \overline{C_2})$ and is determined by:

$$P_{S_2}(t) = p_1(t)p_2(t) + q_1(t) \cdot p_2(t) + p_1(t) \cdot q_2(t) = 1 - q_1(t) \cdot q_2(t),$$

where $p_1(t), p_2(t)$ are the availabilities of $C_1$ and $C_2$, respectively, and $q_1(t), q_2(t)$ are the unavailabilities of $C_1$ and $C_2$, respectively.

Notice that multistate systems always retain this binary nature. It can be described as available or unavailable. The major difference with the elementary component is the presence of the degraded states in the multistate system's availability and unavailability subspaces. While the elementary component does not show any degraded states, it is either available or unavailable, exclusively. It should be recalled that the availability subspace and the unavailability subspace are single-state each in the case of the elementary component.

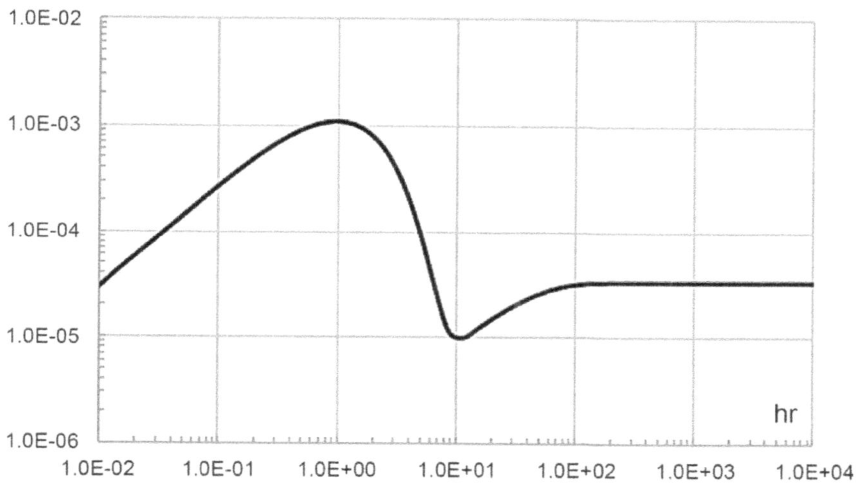


**Figure 2.2.** *The evolution over time of the unavailability of system $S_2$*

The simplest multistate system, with its two unavailability configurations $U_1$and $U_2$, is a system with analytical representation. The probabilities of being in the states and the transition rates between the states are well determined. It can be represented by a binary model whose failure rate and repair rate are well determined at each instant. From now on, we will call them "equivalent failure rate" and "equivalent repair rate". Some authors call them "pseudo-failure" and "pseudo-repair (/recovery)" rates.

Note that regardless of the total number of states in the universal functional space of a multistate system, they can always be decomposed into two disjoint subspaces: availability space and unavailability space. The generalization of the binarization is the subject of the next sections. With binarization, we abstract all degraded states. We only keep the exact kinetics of the system. We should introduce the notion of "critical transitions" before we can develop the model of the equivalent failure and repair rates. From now on, we will be using the terms “availability space”, “unavailability space” and the “functional space” to replace the terms “availability subspace”, “unavailability subspace” and the “universal functional space”.

## 2.2. Critical transitions

We must point out the fact that the multistate system transits from its availability space to its unavailability space via specific transitions called “critical transitions”. We distinguish for the system $S_2$ (Figure 2.1b) two classes of these possible transitions:

– Noncritical transitions: are presented by the transitions in both directions between the states “$E1$ and $E2$” and the states “$E1$ and $E3$”. Whether they are transitions by repair or by failure, the system does not leave its availability space. The availability space consists of a perfect operating state $E1$ and two degraded operating states $E2$ and $E3$.

– Critical transitions: are presented by the transitions between the states “$E2$ and $E4$” and between the states “$E4$ and $E3$”. The system moves back and forth between its availability and its unavailability spaces, depending on the nature of the transition whether it is a failure or a repair action.

An example with a multistate system, whose functional space contains a number of states greater than four, will be discussed in later chapters.

The identification of all the critical transitions in a multistate systems allows the determination of the equivalent parameters $(\lambda^*, \mu^*, \gamma^*)$ of the systems. The binarization of the system is now possible.

## 2.3. Binarization of multistate systems

The binarization of multistate systems having a large number of states reduces the risk of the combinatory explosion of the state graph. A system made up of 10 elementary components would already be described by $2^{10}$ possible states. If we binarize this system by determining its characteristic equivalent parameters $(\lambda^*, \mu^*, \gamma^*)$ at each instant of time, we could treat it as an elementary component with any other binarized big systems, as long as its components are functionally independent and coherent.

Concerning the case of the system $S_2$, we know the states from which the system transits toward the unavailability state $E_4$. Then, we can determine the equivalent failure and repair rates of $S_2$.

We will then determine the equivalent failure and the equivalent repair rates of the system $S_2$. From now on, they are called the "system equivalent transition parameters".

Let $\lambda_{S_2}$ be the equivalent failure rate of $S_2$. The probability that $S_2$ undergoes a transition from its availability space to its unavailability space in the vicinity of the instant $t$ is determined by, see Figure 2.1b:

$$P_{S_2}(t) \cdot \left(\lambda_{S_2} \Delta \mathrm{t}\right) = p_1(t) q_2(t) \cdot (\lambda_1 \Delta t) + q_1(t) p_2(t) \cdot (\lambda_2 \Delta t), \quad [2.1]$$

where:

– $P_{S_2}(t) \cdot \left(\lambda_{S_2} \Delta \mathrm{t}\right)$ is the probability that $S_2$ is available at instant $t$ and then undergoes a transition to the unavailability state around the instant $t$;

– $p_1(t) q_2(t) \cdot (\lambda_1 \Delta t)$ is the probability of being in the state $E_2(C_1, \overline{C_2})$ and then component $C_1$ fails in the interval $\Delta t$;

– $q_1(t) p_2(t) \cdot (\lambda_2 \Delta t)$ is the probability of being in the state $E_3(\overline{C_1}, C_2)$ and then component $C_2$ fails in the interval $\Delta t$;

– $\lambda_{S_2}, \lambda_1, \lambda_2$ are the failure rates of the system $S_2$ and of the components $C_1$ and $C_2$, respectively;

– $p_1(t), p_2(t)$ are the availability of the components $C_1$ and $C_2$, respectively;

– $q_1(t), q_2(t)$ are the unavailability of the components $C_1$ and $C_2$, respectively.

By taking the limit as $\Delta t \to 0$, equation [2.1] can be rewritten in the following form:

$$\lambda_{S_2}(t) = \frac{\lambda_1 \cdot p_1(t) q_2(t) + \lambda_2 \cdot q_1(t) p_2(t)}{P_{S_2}(t)}. \quad [2.2a]$$

Following the same procedure, the system equivalent $\mu_{S_2}(t)$ repair rate will be determined by:

$$\mu_{S_2}(t) = \frac{\mu_1 \cdot q_1(t) q_2(t) + \mu_2 \cdot q_1(t) q_2(t)}{Q_{S_2}(t)} = \mu_1 + \mu_2. \quad [2.2b]$$

Notice that $Q_{S_2}(t)$ is equal to $q_1(t).\,q_2(t)$.

Now, equations [2.2] allow us to validate the model describing the availability $P_{S_2}(t)$ and the unavailability $Q_{S_2}(t)$ of the system $S_2$ using the same model as an elementary component as follows:

$$\frac{d}{dt} P_{S_2}(t) = -\lambda_{S_2}(t) P_{S_2}(t) + \mu_{S_2}(t) Q_{S_2}(t), \quad [2.3a]$$

$$\frac{d}{dt} Q_{S_2}(t) = \lambda_{S_2}(\mathrm{t}).\, p_{S_2}(\mathrm{t}) - \mu_{S_2}(\mathrm{t}).\, q_{S_2}(\mathrm{t}). \quad [2.3b]$$

The evolution of the system equivalent failure rate $\lambda_{S_2}(t)$ over time is plotted in Figure 2.3, using equation [2.2a] and the parameters $(\lambda, \mu, \gamma)$ of the elementary components $C1(10^{-4}\,\mathrm{h}^{-1}, 1\,\mathrm{h}^{-1}, 3 \times 10^{-1})$ and $C2(10^{-2}\,\mathrm{h}^{-1}, 2 \times 10^{-2}\,\mathrm{h}^{-1}, 0)$.

We find that the equivalent rate of system $S_2$ varies with time.

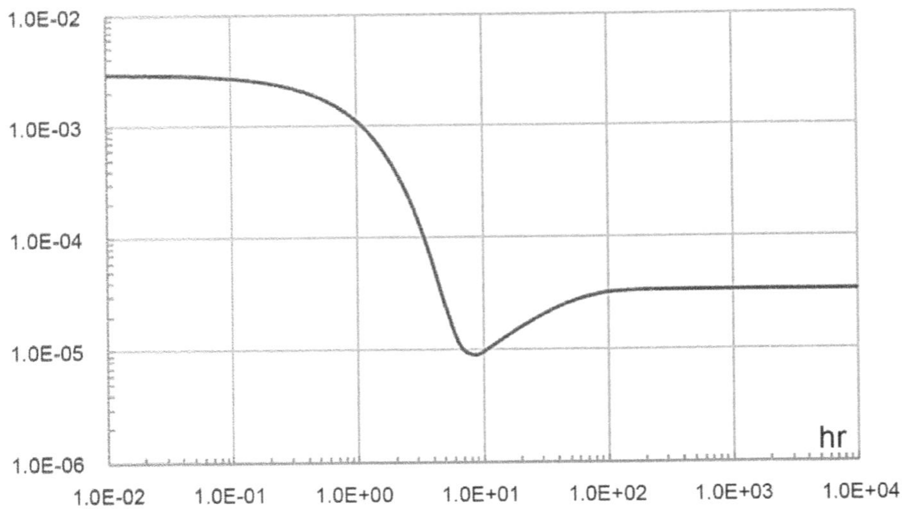


**Figure 2.3.** *Evolution of the equivalent failure rate $\lambda_{S_2}$ over time. For a color version of this figure, see www.iste.co.uk/elhami/reliability.zip*

We can also determine the equivalent repair rate $\mu_{S_2}(t)$ of the system using the same approach (see equation [2.2b]). The equivalent repair rate is constant $\mu_{S_2}(t)$ and equal to 1.2 $h^{-1}$.

EXAMPLE 2.2.–

Let $S_1$ be a system made up of two elementary components $C1$ and $C2$, which operate in series. The system fails if only one of the two components fails. Use the above approach and Figure 2.1a to demonstrate that the equivalent failure rate $\lambda_{S_1}(t)$ of $S_1$ is determined by:

$$\lambda_{S_1}(\mathrm{t}) = \lambda_1 + \lambda_2. \qquad [2.4a]$$

Use the same approach and demonstrate that the equivalent repair rate $\mu_{S_1}$ of $S_1$ is determined by:

$$\mu_{S_1}(\mathrm{t}) = \frac{\mu_1 \cdot q_1(t) p_2(t) + \mu_2 \cdot p_1(t) q_2(t)}{Q_{S_1}(t)}. \qquad [2.4b]$$

Equation [2.4] allows us to validate the model describing the availability $P_{S_1}(t)$ and the unavailability $Q_{S_1}(t)$ of the system $S_1$ using the same model as an elementary component in equations [2.3].

The previous equations [2.3] that described a dynamic model of the availability and unavailability of the elementary binary component would therefore represent a universal dynamic model for binary and multistate coherent systems whose elementary components are independent.

We underline that the approach, in equations [2.2] and [2.4], can be generalized for all systems, similar to $(S1, S2)$, whose availability and unavailability are defined by one of these definitions whatever the number of elementary components:

1) $S1$ : the system is available if only one of the components is available – equations [2.2].

2) $S2$ : the system is unavailable if only one of the components is unavailable – equations [2.4].

Any system made up of $N$ elementary components and belonging to the set $S1$ or $S2$ can be binarized using the previous approach to determine its equivalent parameters ($\lambda^*, \mu^*$). Note that the system $S1$ has only one state in its space of availability space and ($2^N - 1$) states in its unavailability space. Then, it has $N$ possible critical transitions. However, the system $S2$ has only one state in its unavailability space and ($2^N - 1$) states in its availability space. It has $N$ possible critical transitions as well. The generalization of equations [2.2] and [2.4] is direct for the binarization of $S1$- and $S2$-like systems with $N$ elementary components in monotonic structure, either in parallel or in series. In Table 2.1, we summarize the mathematical expressions that describe the equivalent parameters ($\lambda^*, \mu^*$) of $S1$- and $S2$-like systems with $N$ elementary components.

| | **Configuration** | **Equivalent failure (Λ) and repair (*M*)** |
|---|---|---|
| 1 | System $S_1$<br><br>*N* components in series:<br>The system is unavailable if one of its components is unavailable | $\Lambda = \sum_{i=1}^{N} \lambda_i$<br><br>$M = \left[\sum_{i=1}^{N} \mu_i q_i \left(\prod_{\substack{j=1 \\ j \neq i}}^{N} p_j\right)\right] / \left[1 - \left(\prod_{j=1}^{N} p_j\right)\right]$ |
| 2 | System $S_2$<br><br>*N* components in parallel:<br>The system is available if one of its components is available | $\Lambda = \left[\sum_{i=1}^{N} \lambda_i p_i \left(\prod_{\substack{j=1 \\ j \neq i}}^{N} p_j\right)\right] / \left[1 - \left(\prod_{j=1}^{N} q_j\right)\right]$<br><br>$M = \sum_{i=1}^{N} \mu_i$ |

**Table 2.1.** *Equivalent parameters in two specific configurations*

Binarizing a multistate system reduces the number of states from $2^N$ to only two states. Admittedly, we lose the details of the degraded states, but the restitution of the degraded states is simple and straightforward especially in the case of the monotonically structured systems, that is, $S1$- or $S2$-like systems. This will generally not be the case for other complex multistate systems.

To complete the binarization of multistate systems, it remains still to determine the equivalent failure-to-start probability $\gamma^*$ of the system. This will be the subject of section 2.4.

## 2.4. Failure-to-start probability

We remind the reader that the probability of failure-to-start is the initial unavailability $Q_1(0)$ of the system.

If the system is an $S_1$-like system with two elementary components connected in series, its equivalent failure-to-start probability $\gamma_{S_1}$ is determined by:

$$\gamma_{S_1} = Q_1(0) = 1 - (1 - \gamma_1) \cdot (1 - \gamma_2). \quad [2.5a]$$

If the system is an $S_2$-like with two elementary components connected in parallel, its equivalent failure-start probability $\gamma_{S_2}$ is determined by:

$$\gamma_{S_2} = Q_2(0) = \gamma_1 \gamma_2. \quad [2.5b]$$

The generalization of equations [2.5] for systems with $N$ elementary components, either all in series $S_1$ or all in parallel $S_2$, is described by the following equations [2.6]:

$$\gamma_{S_1} = 1 - \prod_{i=1}^{N}(1 - \gamma_i), \quad [2.6a]$$

$$\gamma_{S_2} = \prod_{i=1}^{N} \gamma_i, \quad [2.6b]$$

where $\gamma_S$ and $\gamma_i$ are the failure-to-start probabilities of the systems $(S_1, S_2)$ and the probabilities of failure to start the elementary components of the system.

Before leaving the subject of the binarization of a system with $N$ elementary components arranged in series or in parallel, we must quickly present the notion of the "forbidden transitions". This is necessary to be able to explain why we have only $N$ critical transitions between the $2^N$ states that define the universal functional space of the systems $S_1$ and $S_2$. What about the forbidden transitions?

## 2.5. Forbidden transitions

The transitions between states are classified into critical and non-critical ones. But, to close on the subject concerning the classification of transitions in a multistate system, we must complete it with the presentation of the "forbidden transitions". Then, let us recall that the elementary components are "independent". The independence property of elementary components has a major importance in the identification of the forbidden transitions.

We refer again to Figure 2.1b where transitions between the "$E1$ and $E4$" states are absent. These transitions in both directions (failure and repair) are forbidden, not forgotten. This is because the elementary components involved are, by definition, independent. To demonstrate how the components independence property results in the prohibition of some transitions, we proceed as follows.

Let us suppose that the transition from "$E1$" to "$E4$" (in Figure 2.1b) exists and let $\pi$ be the rate of transition from "$E1$" to "$E4$". The probability $\Delta p$ that the system undergoes this transition in the vicinity of instant $t$ would then be determined by:

$$\Delta p = \pi \cdot \Delta t = (\lambda_1 \cdot \Delta t) \cdot (\lambda_2 \cdot \Delta t),$$

where $(\lambda_1 \cdot \Delta t)$ and $(\lambda_2 \cdot \Delta t)$ are, respectively, the probabilities that $C1$ and $C2$ fail within the time interval $\Delta t$ in the vicinity of the instant $t$. The occurrence probability $\Delta p$ of such double failure is simply the product of the two occurrence probabilities of both independent failures.

This allows us to express the occurrence rate $\pi$ of this double failure event such as:

$$\pi = \frac{\Delta p}{\Delta t} = \frac{\lambda_1 \lambda_2 \cdot [\Delta t]^2}{\Delta t}.$$

By taking the limit when $\Delta t \to 0$ :

$$\lim_{\Delta t \to 0} \pi = \lim_{\Delta t \to 0} \lambda_1 \lambda_2 \cdot \Delta t \to 0.$$

The transition rate $\pi$ cannot be but zero.

The conclusion is that the instantaneous transition of multi-independent events is forbidden. Only dependent events may admit a non-zero instantaneous transition rate. This is, for example, the case of common mode failures.

## 2.6. Boolean models and state graphs equivalence

The previously studied system is the simplest multistate system that can be designed. It is made up of two elementary components. This system can be associated with two possible failure modes depending on the failure's definition. These two possible failure modes are well presented by state graphs in Figure 2.1. One could instead have defined these two failure modes using Boolean models.

For the writing of the Boolean expressions that will follow, we will use the operator "+" for the logical operation "OR", as well as the operator "·" for the logical operation "AND".

The first possible failure configuration was associated with a system $S_1$ that fails if only one of the elementary components fails (Figure 2.1a). The unavailability $U_1$ (or $\overline{A_1}$) of the system will therefore be described by the following Boolean expression:

$$U_1 = \overline{C_1} + \overline{C_2}. \qquad [2.7a]$$

The expression [2.7a] describes the unavailability of the system $S_1$ by the minimal failure conditions, called "minimal cuts". The same expression can be rewritten using maximum cuts, where the state of each elementary component is clearly identified as follows:

$$\begin{aligned} U_1 &= \overline{C_1} \cdot (C_2 + \overline{C_2}) + \overline{C_2} \cdot (C_1 + \overline{C_1}) \\ &= \overline{C_1} \cdot C_2 + C_1 \cdot \overline{C_2} + \overline{C_1} \cdot \overline{C_2} \end{aligned} \qquad [2.7b]$$

We note that expression [2.7b] is the exact Boolean description of the unavailability space as it is defined by the state graph in Figure 2.1a. Expression [2.7a] is only a reduced description of the state graph in Figure 2.1a because it uses the minimal cuts, that is, the minimal system unavailability conditions.

The availability $A_1$ of the system can also be described by the following Boolean expression:

$$A_1 = \overline{U_1} = \overline{(\overline{C_1} + \overline{C_2})} = C_1 \cdot C_2. \qquad [2.7c]$$

The Boolean expression [2.7c] describes the system availability condition by its minimal condition: the two elementary components should be available. The availability space in this case contains a single state $\{C_1 \cdot C_2\}$. The universal functional space $2^2$ of the system $S_1$ contains three states [2.7b] of unavailability and one state [2.7c] of availability.

The same procedure can be repeated to extract the same information from the Boolean expression of the second failure mode that is associated with a system $S_2$. The system $S_2$ would fail if both elementary components fail (Figure 2.1b). The unavailability $U_2$ of the system will therefore be described by the following Boolean expression:

$$U_2 = \overline{C_1} \cdot \overline{C_2}. \quad [2.8a]$$

The availability $A_2$ of the system $S_2$ will then be described by the following Boolean expression:

$$A_2 = \overline{U_2} = \overline{\overline{C_1} \cdot \overline{C_2}} = C_1 + C_2. \quad [2.8b]$$

The system $S_2$ is therefore available if at least one elementary component is available. Expression [2.8b] is formulated by its minimal cuts. This minimal expression hides three logical conjectures. Each represents an availability state in the state graph in Figure 2.1b, in the following form:

$$A_2 = C_1 \cdot C_2 + C_1 \cdot \overline{C_2} + \overline{C_1} \cdot C_2. \quad [2.8c]$$

The conclusion is that each state graph can be expressed by a Boolean expression, whatever the complexity of the multistate system to be described. Boolean expressions can be formulated by minimal or disjoint cuts (maximal). This last possibility will be briefly presented later.

Is the simplest system always the one with two elementary components and four possible functional states ($2^2$)?

Yes, if we deal with systems with analytical structure representation.

But, if we are dealing with systems with systemic representation, the simplest system one can conceive is the one whose universal functional space contains three states.

## 2.7. Systems with systemic structure representation

It also often happens that the functional analysis of a given system does not allow going deep enough into the finest possible granularity to describe the system considering its elementary components (binary and independent). The analyst will then simply stop the analysis at a functional level where the associated success-failure data are available. These functionalities will conventionally be considered basic functionalities. We will not be able to identify what these functionalities may hide or contain as other more basic (sub)functionalities. In this case, the data we need to obtain are systemic data and precisely the functional states definition and the transition rates between these basic functional states. Systemic data are system global data issued through the system operational feedback experience, estimated or through experts' judgement. The major difference between systemic and analytic representation is as follows:

– Using the analytical representation, the analyst can determine the probabilities of being in each state and the transition rates between the states independently. The basic required data are the elementary components parameters $(\lambda, \mu, \gamma)$ and the failure (or success) minimal conditions.

– Using the systemic representation, the analyst cannot determine the probabilities of being in the states independently of the transition rates. the basic required data are the definition of the functional states (the number and the transition rates), the states' initial conditions and the failure (or success) minimal conditions.

It also gives analysts the possibility of reducing the functional space of a multistate system into a smaller number of macro-states (equivalent states) because of the calculations of the equivalent transition rates between these macro-states. In this case, the representation of the system is semi-systemic.

Thus, a multistate system can be perfectly represented by a state graph, based on the knowledge about the macro-states and the transition rates between these macro-states (Figure 2.4). Figure 2.4 represents a simplified epidemiological model which describes the exposed population states with four classes: without symptoms ($E1$), in intensive care ($E2$), with symptoms ($E3$) and deceased ($E0$). The transition rates used are of academic interest only.

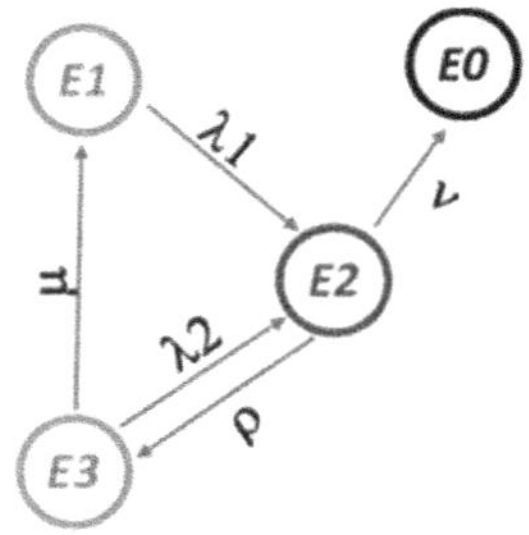


**Figure 2.4.** *A multistate system represented by its state graph. For a color version of this figure, see www.iste.co.uk/elhami/reliability.zip*

The state graph in Figure 2.4 is determined by the following equations:

$$\frac{d}{dt}p_1(t) = -\lambda_1 p_1(t) + 0\, p_2(t) + \mu\, p_3(t) + 0\, p_0(t),$$

$$\frac{d}{dt}p_2(t) = \lambda_1 p_1(t) - (\nu + \rho)p_2(t) + \lambda_2 p_3(t) + 0 p_0(t),$$

$$\frac{d}{dt}p_3(t) = 0 p_1(t) + \rho p_2(t) - (\lambda_2 + \mu)p_3(t) + 0 p_0(t),$$

$$\frac{d}{dt}p_0(t) = 0\, p_1(t) + \nu p_2(t) + 0\, p_3(t) + 0\, p_0(t). \qquad [2.9]$$

The system of differential equations [2.9] can be solved if we have the initial conditions $[p_1(0), p_2(0), p_3(0), p_0(0)]$.

In this example, the state $E_0$ emits no transitions. This is called an "absorbing state". The absorbing state can then be separated from the set of other states because it can be determined, as in this example, by:

$$\frac{d}{dt}p_0(t) = \nu p_2(t). \qquad [2.10]$$

It is therefore sufficient to know the solution of $p_2(t)$ to determine $p_0(t)$.

If the transition rates are time-constant, the state graph (Figure 2.4) would be described as a Markov graph.

## 2.8. The Markov graph

The Markov graph is a state graph under certain conditions. If, and only if, all the transition rates involved in the state graph are constant over time, the system of equations [2.9] will become a linear differential system and will have an analytical solution. It will therefore become a Markovian system. An analytical solution may thus exist.

If certain transition rates involved in the state graph are slow enough over time, the system of differential equations [2.9] will become a semi-Markovian system. An analytical solution could also be developed.

Otherwise, the system of equations [2.9] will be neither Markovian nor semi-Markovian. Only numerical solutions would be possible.

EXAMPLE 2.3.–

Determine the evolution over time of the system described by the Markov graph in Figure 2.4, given the initial conditions [1,0,0,0] and the values of the transition rates in Table 2.2.

| **Rate (per day)** | $\lambda 1$ | $\lambda 2$ | $\nu$ | $\mu$ | $\rho$ |
|---|---|---|---|---|---|
| **($d^{-1}$)** | $7.15^{-1}$ | $3.58^{-5}$ | $2.08^{-3}$ | $2.50^{-3}$ | $5.2^{-2}$ |

**Table 2.2.** *The transition rates in the state graph of Figure 2.4*

In this example, the determination of the probabilities to be in the states $E_1, E_2$ and $E_3$ consists of solving the following system of differential equations:

$$\frac{d}{dt}\begin{bmatrix} p_1 \\ p'_2 \\ p_3 \end{bmatrix} = \begin{pmatrix} \beta_{11} & \beta_{12} & \beta_{13} \\ \beta_{21} & \beta_{22} & \beta_{23} \\ \beta_{31} & \beta_{32} & \beta_{33} \end{pmatrix} \begin{bmatrix} p_1 \\ p'_2 \\ p_3 \end{bmatrix} =$$

$$\begin{pmatrix} -7.15^{-1} & 0 & 2.50^{-3} \\ +7.15^{-1} & -5.2^{-2} & 3.58^{-5} \\ 0 & +5.2^{-2} & -2.54^{-3} \end{pmatrix} \begin{bmatrix} p_1 \\ p'_2 \\ p_3 \end{bmatrix}, \qquad [2.11]$$

where $\beta_{ij}$ and $p_i(t)$ are the transition rates from state $i$ to state $j$ and the probabilities of being in states $i$ ($i = 1,2,3$), respectively. $\beta_{ij}$ is called the transition matrix. The probability $p'_2$ is defined as:

$$p'_2 = p_2 + p_0 = p_2 + p_2 e^{-\nu t} = p_2\,(1 + e^{-\nu t}).$$

The determination of $p'_2$ allows immediately the determination of $p_2$ and of $p_0$.

The solution of the system of differential equations in [2.11] will have the following form:

$$p_i(t) = \sum_{j=1}^{3} c_{ij} e^{\nu_j t}, \qquad [2.12]$$

where, $\nu_j$ are the eigenvalues $(-7.15^{-1}, -5.47^{-2}, -1.0^{-16})$ of the transition matrix in equation [2.11], and the coefficient $c_{ij}$are determined by:

$$\begin{bmatrix} c_{i1} \\ c_{i2} \\ c_{i3} \end{bmatrix} = \begin{pmatrix} 1 & 1 & 1 \\ \nu_1 & \nu_2 & \nu_3 \\ \nu_1^2 & \nu_2^2 & \nu_3^2 \end{pmatrix}^{-1} \begin{pmatrix} \delta_{i1} & \delta_{i2} & \delta_{i3} \\ \beta_{i1} & \beta_{i2} & \beta_{i3} \\ \beta_{i1}^2 & \beta_{i2}^2 & \beta_{i3}^2 \end{pmatrix} \begin{bmatrix} p_1(0) \\ p_2(0) \\ p_3(0) \end{bmatrix}, \qquad [2.13]$$

where $p_i(0)$ are the initial conditions and $\beta_{ij}^2 = \beta_{ik}\beta_{kj}$, $[\ k = 1,2,3]$.

The probabilities of being in states $E_0, E_1, E_2$and $E_3$ are determined by:

$$p_1(t) = +1.001 e^{-7.15^{-1}t} - 3.899^{-3} e^{-5.47^{-2}t} + 3.325^{-3} e^{-1.0^{-16}t},$$

$$p'_2(t) = -1.079 e^{-7.15^{-1}t} + 1.033 e^{-5.47^{-2}t} + 4.634^{-2} e^{-1.0^{-16}t},$$

$$p_3(t) = +7.880^{-2} e^{-7.15^{-1}t} - 1.029 e^{-5.47^{-2}t} + 9.503^{-1} e^{-1.0^{-16}t},$$

and then:

$$p_2(t) = p'_2(t)/(1 + e^{-2.08^{-3}t}),$$

$$p_0(t) = p_2(t).\,e^{-2.08^{-3}t}. \qquad [2.14]$$

The generalization of equation [2.11] to deal with the case of an $n$ −state system is straightforward.

The evolution of the probabilities of being in states $E_0, E_2$ and $E_3$ is plotted in Figure 2.5.

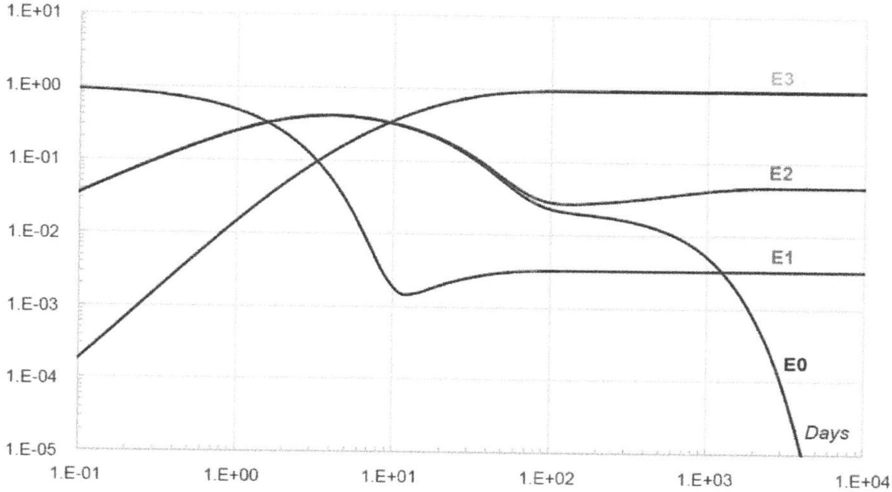


**Figure 2.5.** *The evolution over time of the probabilities of being in states E1 (green), E2 (red), E3 (orange) and E0 (black). For a color version of this figure, see www.iste.co.uk/elhami/reliability.zip*

As we have seen, the determination of the probabilities of being in different states could not have been possible without the knowledge of the number of the states and the transition rates. The entire solution of the problem depends on the a priori knowledge of the transition matrix and the initial condition.

We should highlight how useful the agglomeration of the states $\{E_0, E_2\}$ in one macro-state $E_2'$ was in reducing the number of the states from 4 to 3.

From now on, we will proceed to group the states into macro-states and then determine equivalent transition rates when it is useful. This is often useful if the state graph is voluminous and if the grouping into macro-states allows a more synthetic analysis of the different functional phases of the system.

# 3

# Matrix-like System

## 3.1. Introduction

A matrix-like system, also called a $N \times M$ system, consists of $N$ rows in parallel, where each row consists of $M$ components $C_{nm}$ ($C_{nm}$; $n \in [1, N], m \in [1, M]$) (Figure 3.1). Maximum system performance is achieved when the $N$ lines are fully operational. Maximum system availability is also reached when the $N$ lines are available. No line, in principle, is systematically in standby state. Putting one or more lines on hold could therefore express an incidental or programmed performance degradation. The system is therefore not a system with redundancy, by design.

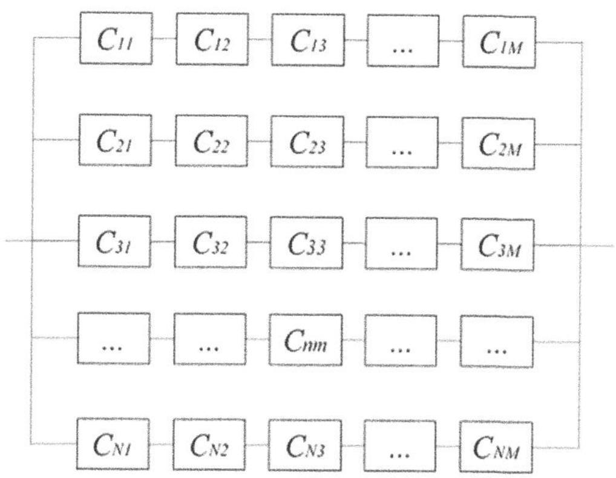


**Figure 3.1.** *Schematic presentation of an $N \times M$ system*

The components $C_{nm}$ $(C_{nm};\ n \in [1, N], m \in [1, M])$ are generally different, independent and coherent. We will not deal with the case of dependent and non-coherent components.

Matrix systems are generally used for the transport/transmission of matter, energy or information between a source and a recipient. Their performance is often measured by a metric appropriate to the transported/transmitted medium. The unavailability, by its probabilistic metric, contributes to the evaluation of the global performance of the system. It serves as a weighting factor more than an overall performance index, for example, in supply chains performance assessments.

Consequently, the analyst focuses instead on the availability of the lines, line by line. Describing the functional state of the system by the number of the available lines is an objective priority. It is much more meaningful than the overall availability of the system.

In this section, we are thus interested in modeling the performance of a general $N \times M$ system and its dynamics.

## 3.2. Modeling the functional state of the system

The modeling of the functional state of the system is done in two stages.

– First, we try to binarize a line, which is made up of $M$ independent components. The components are generally different, and each is well defined by its characteristic parameters $(\lambda, \mu, \gamma)$. Binarization of the line is accomplished by determining the equivalent characteristic parameters of the line $(\lambda_n, \mu_n, \gamma_n)$, where the index $n$ is the line identifier. From now on, all the components of the line $n$ will be represented by a single entity, called a module, whose characteristic parameters are $(\lambda_n, \mu_n, \gamma_n)$.

– In the second step, we classify the possible functional states of the system in sets $\mathcal{L}_l^N$ $\{\mathcal{L}_l^N; l = 0, 1, \dots, N\}$, where $l$ and $N$ are, respectively, the number of unavailable modules and the total number of modules in the $N \times M$ system. We will then focus on the probability of sojourn in a set of states and on the transition rates between the states.

### 3.2.1. *Modeling a line of elementary components*

Considering the independence of the elementary components, the availability $P_n(t)$ of the $n^{th}$ line is determined by:

$$P_n(t) = \prod_{m=1}^{M} p_{nm}(t), \quad [3.1a]$$

where $p_{nm}(t)$ is the availability of the component $m$ in the row $n$.

Subsequently, the unavailability $Q_n(t)$ of the $n^{th}$ line is determined by:

$$Q_n(t) = \left(1 - \prod_{m=1}^{M} p_{nm}(t)\right). \quad [3.1b]$$

The equivalent failure rate $\lambda_n(t)$ of the line $n$ is determined by:

$$\lambda_n(t) = \sum_{m=1}^{M} \lambda_{nm}(t), \quad [3.2]$$

where $\lambda_{nm}(t)$ is the failure rate of the component $m$ in the line $n$, at instant "$t$".

The equivalent repair rate $\mu_n(t)$ of the line $n$ is determined by:

$$\mu_n(t) = \sum_{j=1}^{M} \left( \frac{\mu_{nj}(t) q_{nj}(t)}{Q_n(t)} \prod_{\substack{m=1 \\ m \neq j}}^{M-1} p_{nm}(t) \right), \quad [3.3]$$

where $\mu_{nm}(t)$ is the repair rate of the component $m$ in the line $n$, at instant "$t$".

From now on, lines can be considered as $N$ elementary and binary components, because of equations [3.1–3.3]. We will therefore replace the term "line" by "module". A module is now fully defined by its characteristic parameters $(\lambda_n, \mu_n, \gamma_n)$ at each instant '$t$'

### 3.2.2. *System functional space modeling*

Having binarized the lines and having renamed them "modules", we can henceforth construct the functional space of the system using a state graph made up of $2^N$ states. Each state is defined by a binary state structure function. A binary state structure function has a binary value and [0,1] and is dependent on the state binary functions [0,1] of all the modules in the systems. The module binary state function has the value 0 if it is available; otherwise, the value is 1. That is the same for the system binary state structure function. We propose the following simple case to illustrate the binary state structure function.

Let 0 and 1 be the values of the binary state structure function of a module to indicate its availability and its unavailability, respectively. Consider the functional space of a system made up of five independent modules and its 32 states ($2^5$). If we arrange these states in ascending binary order because of their binary structure functions, we can immediately determine the number of the unavailable modules in each of these 32 states (Table 3.1).

The $2^N$ states can therefore be classified into sets of states $\mathcal{L}_l^N$ ($\mathcal{L}_l^N; l = 0,1,2,\dots,N$) according to the exact number of $l$ unavailable modules. Each set $\mathcal{L}_l^N$ contains $k_l$ states with $l$ unavailable modules. The number of states $k_l$ is determined by:

$$k_l = \binom{N}{l} = \frac{N!}{l!(N-l)!}. \qquad [3.4]$$

In accordance with equation [3.4], we find that there is one state with 0 unavailable modules ($l = 0$), five states with one unavailable module ($l = 1$), 10 states with two unavailable modules ($l = 2$) and so forth.

To simplify the formal expressions of the probabilities and transition rates to follow, we will opt for a system of relative renumbering for the designation of the modules in the states and the designation of the states in the sets of states $\mathcal{L}_l^N$.

For example, if we are interested in the set $\mathcal{L}_3^5$( $k_3 = \binom{5}{3} = 10$) (Table 3.1), we can identify the 10 states, which have three unavailable modules, such as (00111), (01011), (01101), (01110), (10011), (10101), (10110), (11001), (11010) and (11100). If we replace the structure function of these states by the names of the unavailable module, we will have $(m_3, m_4, m_5)$, $(m_2, m_4, m_5)$, $(m_2, m_3, m_5)$, $(m_2, m_3, m_4)$, $(m_1, m_4, m_5)$, $(m_1, m_3, m_5)$, $(m_1, m_3, m_4)$, $(m_1, m_2, m_5)$, $(m_1, m_2, m_4)$ and $(m_1, m_2, m_3)$, where $m_i$ ($m_i; i = 1,2,3,4,5$) designates the $i$th module.

| # | $m1$ | $m2$ | $m3$ | $m4$ | $m5$ | $l$ | Unavailable modules |
|---|---|---|---|---|---|---|---|
| **1** | 0 | 0 | 0 | 0 | 0 | 0 | – |
| **2** | 0 | 0 | 0 | 0 | 1 | 1 | (5) |
| **3** | 0 | 0 | 0 | 1 | 0 | 1 | (4) |
| **4** | 0 | 0 | 0 | 1 | 1 | 2 | (4,5) |
| **5** | 0 | 0 | 1 | 0 | 0 | 1 | (3) |
| **6** | 0 | 0 | 1 | 0 | 1 | 2 | (3,5) |
| **7** | 0 | 0 | 1 | 1 | 0 | 2 | (3,4) |
| **8** | 0 | 0 | 1 | 1 | 1 | 3 | (3,4,5) |
| **9** | 0 | 1 | 0 | 0 | 0 | 1 | (2) |
| **10** | 0 | 1 | 0 | 0 | 1 | 2 | (2,5) |
| **11** | 0 | 1 | 0 | 1 | 0 | 2 | (2,4) |
| **12** | 0 | 1 | 0 | 1 | 1 | 3 | (2,4,5) |
| **13** | 0 | 1 | 1 | 0 | 0 | 2 | (2,3) |
| **14** | 0 | 1 | 1 | 0 | 1 | 3 | (2,3,5) |
| **15** | 0 | 1 | 1 | 1 | 0 | 3 | (2,3,4) |
| **16** | 0 | 1 | 1 | 1 | 1 | 4 | (2,3,4,5) |
| **17** | 1 | 0 | 0 | 0 | 0 | 1 | (1) |
| **18** | 1 | 0 | 0 | 0 | 1 | 2 | (1,5) |
| **19** | 1 | 0 | 0 | 1 | 0 | 2 | (1,4) |
| **20** | 1 | 0 | 0 | 1 | 1 | 3 | (1,4,5) |
| **21** | 1 | 0 | 1 | 0 | 0 | 2 | (1,3) |
| **22** | 1 | 0 | 1 | 0 | 1 | 3 | (1,3,5) |
| **23** | 1 | 0 | 1 | 1 | 0 | 3 | (1,3,4) |
| **24** | 1 | 0 | 1 | 1 | 1 | 4 | (1,3,4,5) |
| **25** | 1 | 1 | 0 | 0 | 0 | 2 | (1,2) |
| **26** | 1 | 1 | 0 | 0 | 1 | 3 | (1,2,5) |
| **27** | 1 | 1 | 0 | 1 | 0 | 3 | (1,2,4) |
| **28** | 1 | 1 | 0 | 1 | 1 | 4 | (1,2,4,5) |
| **29** | 1 | 1 | 1 | 0 | 0 | 3 | (1,2,3) |
| **30** | 1 | 1 | 1 | 0 | 1 | 4 | (1,2,3,5) |
| **31** | 1 | 1 | 1 | 1 | 0 | 4 | (1,2,3,4) |
| **32** | 1 | 1 | 1 | 1 | 1 | 5 | (1,2,3,4,5) |

**Table 3.1.** *Binary state structure functions*

If we renumber in each state the unavailable modules from left to right ($j_k$), the module $m_2$ would be number 1 in states 2, 3 and 4 ($j_2 = j_3 = j_4 = 1$) and would be number 2 in states 8, 9 and 10 ( $j_8 = j_9 = j_{10} = 2$). Modules in these states will therefore be renumbered from one to 10 using the relative index $j_k$ each time one changes the state $k$ in the set $\mathcal{L}_l^N$.

This example illustrates the one-to-one mapping relationship between the binary state structure function and the states of the basic components in each set $\mathcal{L}_l^N$ $\{\mathcal{L}_l^N; l \in [1, N]\}$.

Let the unavailable modules be indexed by the symbol $j$ and the available modules by the symbol $i$. The probability $W_l(t)$ that the system is in the set $\mathcal{L}_l^N$ with $l$ unavailable modules can be determined by:

$$W_l(t) = \sum_{k=1}^{k_l}\left(\prod_{j_k=1}^{l} Q_{j_k}(t) \cdot \prod_{\substack{i_k=1 \\ i_k \neq j_k}}^{N} P_{i_k}(t)\right), \quad [3.5]$$

where $W_0(t) = \prod_{i_k=1}^{N} P_{i_k}(t)$, $W_N(t) = \prod_{j_k=1}^{N} Q_{j_k}(t)$ and:

– the indices $j_k$ and $i_k$, respectively, identify the unavailable and available modules in the $k^{th}$ belonging to the set of states $\mathcal{L}_l^N$;

– the index $k_l$ is determined by equation [3.4]. $P_k(t)$ and $Q_k(t)$ are determined by equation [3.1];

– the states belonging to each set $\mathcal{L}_l^N$ are renumbered with the index $j_k$, which varies from one to $l$, where $l$ indicates the number of the unavailable modules (see Table 3.2).

At this stage of development, it seems useful to us to present a simple numerical application. Consider the case of an $N \times M$ matrix-like system, with $N$ lines in parallel and each line consists of $M$ components $C_{nm}$ in series, $\{C_{nm};\ n \in [1,10], m \in [1,5]\}$ (Figure 3.1). The 50 components $C_{nm}$ are independent and identical. The characteristic parameters $(\lambda, \mu, \gamma)$ of a component are $(2 \times 10^{-3}\text{h}^{-1}, 10^{-1}\text{h}^{-1}, 0)$.

The binarization of each line gives 10 identical modules $m_i$ $\{m_i; (m_i; i = 1,2, \dots, 10)\}$ in parallel. The characteristic parameters of each module $m_i$ are determined by equations [3.1–3.3]. The evolution over time of the unavailability of the module $m_i$ is plotted in Figure 3.2.

| # | $m1$ | $m2$ | $m3$ | $m4$ | $m5$ | $l$ | $j_k$ |
|---|---|---|---|---|---|---|---|
| **1** | 0 | 0 | 0 | 0 | 0 | 0 | 1 |
| **2** | 0 | 0 | 0 | 0 | 1 | 1 | 1 |
| **3** | 0 | 0 | 0 | 1 | 0 | 1 | 2 |
| **4** | 0 | 0 | 0 | 1 | 1 | 2 | 1 |
| **5** | 0 | 0 | 1 | 0 | 0 | 1 | 3 |
| **6** | 0 | 0 | 1 | 0 | 1 | 2 | 2 |
| **7** | 0 | 0 | 1 | 1 | 0 | 2 | 3 |
| **8** | 0 | 0 | 1 | 1 | 1 | 3 | 1 |
| **9** | 0 | 1 | 0 | 0 | 0 | 1 | 4 |
| **10** | 0 | 1 | 0 | 0 | 1 | 2 | 4 |
| **11** | 0 | 1 | 0 | 1 | 0 | 2 | 5 |
| **12** | 0 | 1 | 0 | 1 | 1 | 3 | 2 |
| **13** | 0 | 1 | 1 | 0 | 0 | 2 | 6 |
| **14** | 0 | 1 | 1 | 0 | 1 | 3 | 3 |
| **15** | 0 | 1 | 1 | 1 | 0 | 3 | 4 |
| **16** | 0 | 1 | 1 | 1 | 1 | 4 | 1 |
| **17** | 1 | 0 | 0 | 0 | 0 | 1 | 5 |
| **18** | 1 | 0 | 0 | 0 | 1 | 2 | 7 |
| **19** | 1 | 0 | 0 | 1 | 0 | 2 | 8 |
| **20** | 1 | 0 | 0 | 1 | 1 | 3 | 5 |
| **21** | 1 | 0 | 1 | 0 | 0 | 2 | 9 |
| **22** | 1 | 0 | 1 | 0 | 1 | 3 | 6 |
| **23** | 1 | 0 | 1 | 1 | 0 | 3 | 7 |
| **24** | 1 | 0 | 1 | 1 | 1 | 4 | 2 |
| **25** | 1 | 1 | 0 | 0 | 0 | 2 | 10 |
| **26** | 1 | 1 | 0 | 0 | 1 | 3 | 8 |
| **27** | 1 | 1 | 0 | 1 | 0 | 3 | 9 |
| **28** | 1 | 1 | 0 | 1 | 1 | 4 | 3 |
| **29** | 1 | 1 | 1 | 0 | 0 | 3 | 10 |
| **30** | 1 | 1 | 1 | 0 | 1 | 4 | 4 |
| **31** | 1 | 1 | 1 | 1 | 0 | 4 | 5 |
| **32** | 1 | 1 | 1 | 1 | 1 | 5 | 1 |

**Table 3.2.** *Relative ranking $k_l$ of states according to the number of the unavailable modules $l$*

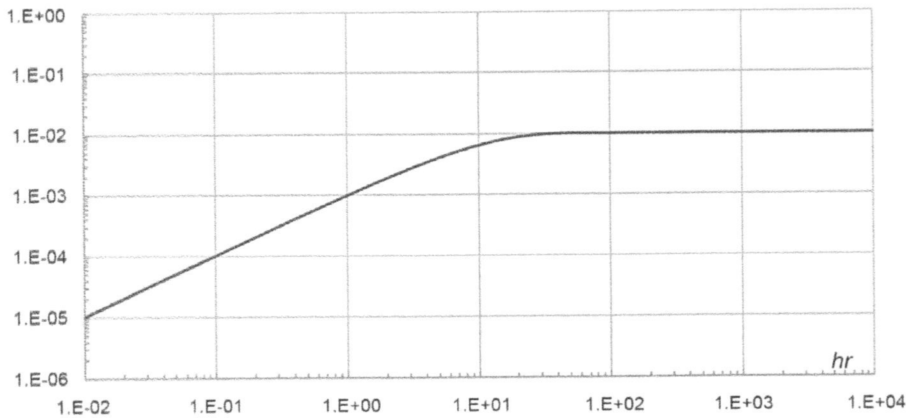


**Figure 3.2.** *The evolution over time of the unavailability of the module* $m_i$*. For a color version of this figure, see www.iste.co.uk/elhami/reliability.zip*

The functional space of the system contains $2^{10}$ binary-structured states. We are interested in the probability $W_l(t)$ of being in the set $\mathcal{L}_l^5$ states at exactly $l$ unavailable modules (equation [3.5]). The evolutions in time of the probability $W_l(t)$ for all the possible values of $l$, $(l = 0,1,2,3,4,5)$, are illustrated in Figure 3.3. The evolution over time of the probability $W_{l=0}(t)$ of being in the state where no module is unavailable, relative to the set $\mathcal{L}_0^5$, is also plotted alone in Figure 3.4.

It is worth remembering that:

$$\sum_{l=0}^{N} W_l(t) \ = 1.0,$$

at every instant $t$.

Once the probabilities $W_l(t)$ of all the sets $\mathcal{L}_l^N$ $(\mathcal{L}_l^N; l = 0,1,2, \dots, N)$ are determined, we can proceed to the determination of the probability $W_\Omega(t)$ that the $N \times M$ system is in a given subspace $\Omega$ defined by the union of certain well-defined sets $\mathcal{L}_l^N$.

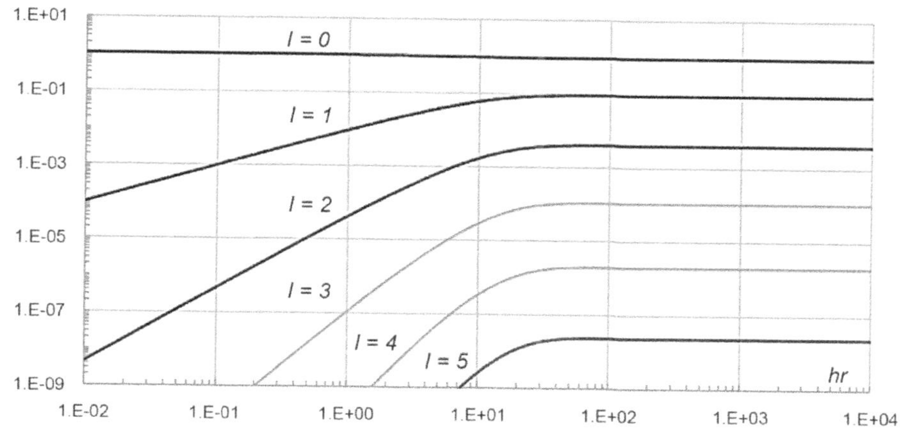


**Figure 3.3.** *The evolution over time of the probability $W_l(t), (l = 0,1,2,3,4,5)$. For a color version of this figure, see www.iste.co.uk/elhami/reliability.zip*

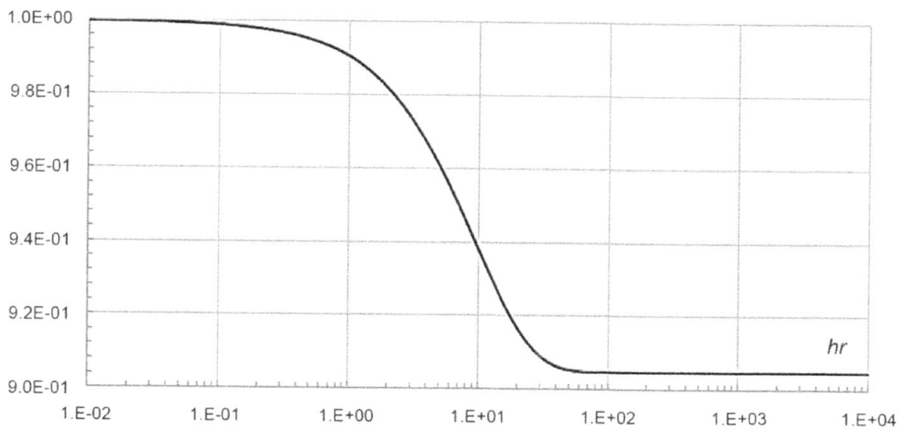


**Figure 3.4.** *The evolution over time of the probability $W_{l=0}(t)$*

### 3.2.3. *The subspaces of sets*

We are interested, here, in the determination of the probability $W_\Omega(t)$ that the $N \times M$ system is in a subspace $\Omega$ defined by the union of certain sets $\mathcal{L}_l^N$, such as:

$$\Omega \equiv \bigcup_{l=l_1}^{l_2} \mathcal{L}_l^N \ ; 1 \leq l_1 \leq l_2 \leq N. \tag{3.6}$$

We recall that the sets $\mathcal{L}_l^N$ are mutually exclusive:

$$\bigcap_l^k \mathcal{L}_l^N = \emptyset \, ; 1 \leq l \neq k \leq N. \tag{3.7}$$

Then, the probability $W_\Omega(t)$ that the $N \times M$ system is in the subspace $\Omega$ is determined by the sum of all the $W_l(t)$ of the sets $\mathcal{L}_l^N$ included in the subspace $\Omega$:

$$W_\Omega(t) = \sum_{l_1}^{l_2} W_l(t). \tag{3.8}$$

Quite often in system reliability, the condition of the grouping of states is expressed by a minimum condition, such as "at least $l_{\min}$ modules are unavailable among $N$". In this case, equation [3.8] would take the following form:

$$W_\Omega(t) = \sum_{l_{\min}}^{N} W_l(t). \tag{3.9a}$$

However, the condition of the grouping of states could also be expressed by a maximum condition, such as "at most $l_{\max}$ modules are unavailable among $N$". In this case, equation [3.8] would take the following form:

$$W_\Omega(t) = \sum_{1}^{l_{\max}} W_l(t). \tag{3.9b}$$

It is interesting to underline that equations [3.9] establish the formal link between the system performance analysis standpoint of view and the system availability-reliability one. The first standpoint of view is interested in the different operating states and their possibilities of combination into sets under different criterion types, without excluding the availability-reliability criteria based on minimum operational conditions. The second standpoint focuses essentially and a priori on the minimal conditions of the unavailability and the reliability of systems.

One of the possible criteria for classifying the functional states of a system is that of the "number $l$ of unavailable modules per state", presented in this chapter. Its interest is of a practical nature. It takes its reasons from the real functional world of systems. It is generally observed that "higher is the number of the unavailable modules in a system, lower is the probability of supplying a required demand over an expected interval of time". Let us take a numerical application for demonstration.

In the case of the $10 \times 5$ matrix system the probability $W_l(t)$ is plotted in Figure 3.3 and the probability $W_\Omega(t)$ is plotted in Figure 3.5 for the minimal unavailability conditions ($l_{\min}$; $l_{\min} = 1,2,3,4,5$) (equation [3.9]).

The numerical differences between the profiles in Figures 3.3 and 3.5 are very small. This is because a differential of almost two decades separating each two successive curves relative to two successive $l$'s.

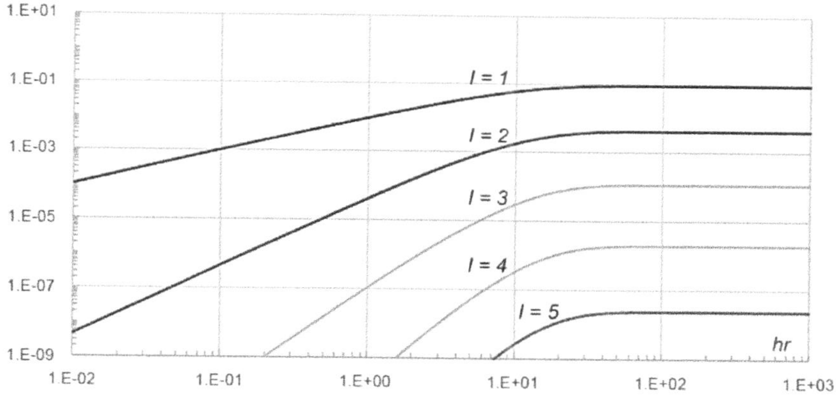


**Figure 3.5.** *The evolution over time of the unavailability of the system $Q_{sys}(t)$, ($l_{min}$ = 1,2,3,4,5). For a color version of this figure, see www.iste.co.uk/elhami/reliability.zip*

To complete our knowledge of the classification of states into sets $\mathcal{L}_l^N$, it is now quite natural to take an interest in the transitions between sets $\mathcal{L}_l^N$.

### 3.2.4. *Transitions between sets*

Being defined by the exact number of unavailable modules $l$, the sets $\mathcal{L}_l^N$, ($\mathcal{L}_l^N; l = 0,1,2, \ldots, N$), only consist of critical states, at the same time. That is to say that the repair of one of the unavailable modules $l$ causes the transit of the functional state of the system from the set $\mathcal{L}_l^N$ to the set $\mathcal{L}_{l-1}^N$ and that the failure of one of the available modules ( $N - l$) causes the transit of the functional state of the system of the set $\mathcal{L}_l^N$ to the set $\mathcal{L}_{l+1}^N$. Note that transitions are possible either to the set $\mathcal{L}_{l-1}^N$ or to the set $\mathcal{L}_{l+1}^N$exclusively. A transition to the set $\mathcal{L}_{l-1}^N$ is considered a recovery of the system performance. However, a transition to the set $\mathcal{L}_{l+1}^N$ is considered a degradation of the system performance. Transitions of order greater than 1 are forbidden because of the functional independence assumed between the modules.

### 3.2.5. *Modeling of equivalent rates of transitions*

Once the critical states are identified, the equivalent transition rates between the sets $\mathcal{L}_l^N$ are immediately determined. We recall that the equivalent rates of transitions are a generic definition to designate the equivalent rate of degradation and the equivalent rate of recovery of the system.

The binarization of a multistate system is generally achieved by determining its equivalent rates of transitions between states or between sets of states.

Let us take the previous case of a system with five modules and seek to determine the equivalent rate of degradation from the set $\mathcal{L}_3^5$ ( $\mathcal{L}_3^5 \to \mathcal{L}_4^5$). The set $\mathcal{L}_3^5$ contains the following states: (00111), (01011), (01101), (01110), (10011), (10101), (10110), (11001), (11010) and(11100). Considering these critical states in the same ranking order, the equivalent rate of degradation $\Lambda_3(t)$ can be determined by:

$$\begin{aligned}
&\Lambda_3(t) \cdot W_3(t) = \\
&+(\lambda_1 + \lambda_2) \cdot P_1(t)P_2(t)Q_3(t)Q_4(t)Q_5(t) \\
&+(\lambda_1 + \lambda_3) \cdot P_1(t)Q_2(t)P_3(t)Q_4(t)Q_5(t) \\
&+(\lambda_1 + \lambda_4) \cdot P_1(t)Q_2(t)Q_3(t)P_4(t)Q_5(t) \\
&+(\lambda_1 + \lambda_5) \cdot P_1(t)Q_2(t)Q_3(t)Q_4(t)P_5(t) \\
&+(\lambda_2 + \lambda_3) \cdot Q_1(t)P_2(t)P_3(t)Q_4(t)Q_5(t) \\
&+(\lambda_2 + \lambda_4) \cdot Q_1(t)P_2(t)Q_3(t)P_4(t)Q_5(t) \\
&+(\lambda_2 + \lambda_5) \cdot Q_1(t)P_2(t)Q_3(t)Q_4(t)P_5(t) \\
&+(\lambda_3 + \lambda_4) \cdot Q_1(t)Q_2(t)P_3(t)P_4(t)Q_5(t) \\
&+(\lambda_3 + \lambda_5) \cdot Q_1(t)Q_2(t)P_3(t)Q_4(t)P_5(t) \\
&+(\lambda_4 + \lambda_5) \cdot Q_1(t)Q_2(t)Q_3(t)P_4(t)P_5(t),
\end{aligned} \quad [3.10]$$

where $P_n(t)$ and $Q_n(t)$ are determined by equations [3.1a] and [3.1b], respectively. The general expression for the probability $W_l(t)$ is described by equation [3.5].

The same approach is used to determine the equivalent recovery rate $M_3(t)$ from the set $\mathcal{L}_3^5$ ( $\mathcal{L}_3^5 \rightarrow \mathcal{L}_2^5$), relying on the same reasoning scheme. This recovery rate $M_3(t)$ is expressed by:

$$\begin{aligned}
&M_3(t) \cdot W_3(t) = \\
&+(\mu_3+\mu_4+\mu_5) \cdot P_1(t)P_2(t)Q_3(t)Q_4(t)Q_5(t) \\
&+(\mu_2+\mu_4+\mu_5) \cdot P_1(t)Q_2(t)P_3(t)Q_4(t)Q_5(t) \\
&+(\mu_2+\mu_3+\mu_5) \cdot P_1(t)Q_2(t)Q_3(t)P_4(t)Q_5(t) \\
&+(\mu_2+\mu_3+\mu_4) \cdot P_1(t)Q_2(t)Q_3(t)Q_4(t)P_5(t) \\
&+(\mu_1+\mu_4+\mu_5) \cdot Q_1(t)P_2(t)P_3(t)Q_4(t)Q_5(t) \\
&+(\mu_1+\mu_3+\mu_5) \cdot Q_1(t)P_2(t)Q_3(t)P_4(t)Q_5(t) \\
&+(\mu_1+\mu_3+\mu_4) \cdot Q_1(t)P_2(t)Q_3(t)Q_4(t)P_5(t) \\
&+(\mu_1+\mu_2+\mu_5) \cdot Q_1(t)Q_2(t)P_3(t)P_4(t)Q_5(t) \\
&+(\mu_1+\mu_2+\mu_4) \cdot Q_1(t)Q_2(t)P_3(t)Q_4(t)P_5(t) \\
&+(\mu_1+\mu_2+\mu_3) \cdot Q_1(t)Q_2(t)Q_3(t)P_4(t)P_5(t),
\end{aligned} \qquad [3.11]$$

where $P_n(t)$ and $Q_n(t)$ are, respectively, determined by equations [3.1a] and [3.1b]. The probability $W_3(t)$ of being in the functional space $\mathcal{L}_3^5$ is determined by:

$$\begin{aligned}
&W_3(t) = \\
&+P_1(t)P_2(t)Q_3(t)Q_4(t)Q_5(t) \\
&+P_1(t)Q_2(t)P_3(t)Q_4(t)Q_5(t) \\
&+P_1(t)Q_2(t)Q_3(t)P_4(t)Q_5(t) \\
&+P_1(t)Q_2(t)Q_3(t)Q_4(t)P_5(t) \\
&+Q_1(t)P_2(t)P_3(t)Q_4(t)Q_5(t) \\
&+Q_1(t)P_2(t)Q_3(t)P_4(t)Q_5(t) \\
&+Q_1(t)P_2(t)Q_3(t)Q_4(t)P_5(t) \\
&+Q_1(t)Q_2(t)P_3(t)P_4(t)Q_5(t) \\
&+Q_1(t)Q_2(t)P_3(t)Q_4(t)P_5(t) \\
&+Q_1(t)Q_2(t)Q_3(t)P_4(t)P_5(t).
\end{aligned} \qquad [3.12]$$

The transitions between the sets $\mathcal{L}_l^N$ are schematically presented in Figure 3.6.

To generalize equations [3.10–3.12], we will renumber the unavailable modules with the index $j$ $\{j; j = 1,2, \dots, l\}$ and the available modules with the index $\{i; i = 1,2, \dots, (N-l)\}$.

Then, we can determine the equivalent transition rates of degradation $\lambda_l(t)$ by:

$$\Lambda_l(t) \cdot W_l(t) = \sum_{k=1}^{k_l}\left[\left(\sum_{i_k=1}^{N-l} \lambda_{i_k}(t)\right)\left(\prod_{i_k=1}^{N-l} P_{i_k}(t)\right)\left(\prod_{j_k=1}^{l} Q_{j_k}(t)\right)\right]. \quad [3.13]$$

The equivalent recovery rate $\mu_l(t)$ will be determined by:

$$M_l(t) \cdot W_l(t) = \sum_{k=1}^{k_l}\left[\left(\sum_{j_k=1}^{l} \mu_{j_k}(t)\right)\left(\prod_{i_k=1}^{N-l} P_{i_k}(t)\right)\left(\prod_{j_k=1}^{l} Q_{j_k}(t)\right)\right] [3.14],$$

where:

$$W_l(t) = \sum_{k=1}^{k_l}\left[\left(\prod_{i_k=1}^{N-l} P_{i_k}(t)\right)\left(\prod_{j_k=1}^{l} Q_{j_k}(t)\right)\right] \quad [3.15]$$

and $k_l$ is the number of states belonging to the set $\mathcal{L}_l^N$ and is determined by:

$$k_l = \binom{N}{l}.$$

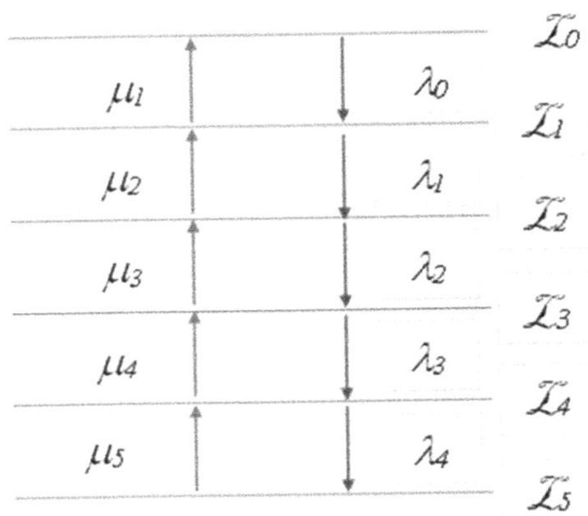


**Figure 3.6.** *Schematic presentation of transitions between sets of states $\mathcal{L}_l^N$ ($\mathcal{L}_l^N; l = 0,1,2, \dots, N$). For a color version of this figure, see www.iste.co.uk/elhami/reliability.zip*

This approach allows us to represent a set of states by a single macro-state, clearly determined by its two transition rates (recovery and degradation) and by an initial condition. The sets $\mathcal{L}_l^N$ are perfectly binarized and determined by their transition rates $(\Lambda_n(t), M_n(t), \Gamma_n)$, where $\Gamma_n = W_n(0)$. We can therefore proceed to the generalization of this dynamic model that describes the transitions between the sets $\mathcal{L}_l^N$ (Figure 3.6).

### 3.2.6. *Dynamic modeling of operating states*

Having described the functional space of the matrix-like system by its sets $\mathcal{L}_l^N$ (Figure 3.6), one can determine the performance of the system using the following system of differential equations:

$$\frac{d}{dt}W_0(t) = (0) - (\Lambda_0 + 0)W_0(t) + M_1 W_1(t),$$

$$\frac{d}{dt}W_1(t) = +\Lambda_0 W_0(t) - (\Lambda_1 + M_1)W_1(t) + M_2 W_2(t),$$

$$\frac{d}{dt}W_2(t) = +\Lambda_1 W_1(t) - (\Lambda_2 + M_2)W_2(t) + M_3 W_3(t),$$

::::::::::::::::::::::::::::::::::::::::::::::::::::::::::::::::::::::::::::::::::::::::::::::::::::

$$\frac{d}{dt}W_N(t) = +\Lambda_{N-1}W_{N-1}(t) - (0 + M_N)W_N(t) + (0). \qquad [3.16]$$

The transitions between the sets $\mathcal{L}_l^N$ are governed by a linear system of first-order differential equations of the form:

$$\frac{d}{dt}W_l(t) = \sum_{j=0}^{N} \alpha_{lj} W_j(t), \qquad [3.17]$$

where:

$$\underset{l \neq k}{\alpha_{lk}} = \begin{cases} \Lambda_{l-1}, & \text{if } k = l-1 \\ M_{l+1}, & \text{if } k = l+1 \\ 0 & \text{if otherwise} \end{cases}$$

$$\alpha_{ll} = -(\Lambda_l + M_l), \qquad M_0 = \Lambda_N = 0, \qquad l = 0, 1, 2, \ldots, N.$$

Equations [3.17] can also be expressed using the following form:

$$\frac{d}{dt}\begin{bmatrix} W_0(t) \\ W_1(t) \\ W_2(t) \\ \dots\dots\dots \\ W_{n-1}(t) \\ W_n(t) \end{bmatrix} = \begin{pmatrix} -\Lambda_0 & M_1 & & & 0 \quad 0 \quad 0 \\ \Lambda_0 & -(\Lambda_1 + M_1) & M_2 & & 0 \quad 0 \quad 0 \\ 0 & \Lambda_1 & -(\Lambda_2 + M_2) & M_3 & 0 \quad 0 \\ & & & & \\ 0 \quad 0 \quad 0 & & \Lambda_{n-2} & -(\Lambda_{n-1} + M_{n-1}) & M_n \\ 0 \quad 0 \quad 0 & & & 0 \quad \Lambda_{n-1} & -M_n \end{pmatrix} \begin{bmatrix} W_0(t) \\ W_1(t) \\ W_2(t) \\ \dots\dots\dots \\ W_{n-1}(t) \\ W_n(t) \end{bmatrix}$$

[3.18a]

This dynamic model characterizes systems where transitions between states are sequential, that is, starting from a state $N$, the system can only have transitions to the states $(N-1)$ or $(N+1)$ (Figure 3.6).

In a matrix-like system, if the components that constitute each module are well defined by their characteristic parameters $(\lambda, \mu, \gamma)$, the determination of the probability $W_l(t)$ of being in a set of states with $l$ unavailable components is straightforward. Transition rates $(\Lambda_n(t), \mathrm{M}_n(t))$ and the probabilities $W_l(t)$ will be determined by equations [3.13–3.15].

If the components are not defined, but the functional sets $\mathcal{L}_n^N$ of the system are well defined by their transition rates $(\Lambda_n(t), M_n(t))$, and their initial conditions, $\Gamma_n = W_n(0)$, we will use the model described by equation [3.17] to determine the probabilities $W_l(t)$.

A particularly interesting subclass of the matrix-like system is the homogeneous $N \times M$ system.

## 3.3. Homogeneous matrix-like system

The components that constitute the modules of a homogeneous $N \times M$ system are independent and identical, that is, they have the same characteristic parameters $(\lambda, \mu, \gamma)$. Therefore, the availability $p(t)$ and the unavailability $q(t)$ are well determined and identical for all components.

As a result, the expressions describing the previously developed models can be significantly simplified.

The availability $P(t)$ of a module made up of $M$ independent and identical components will be determined by:

$$P(t) = \prod_{m=1}^{M} p(t) = p^M(t), \qquad [3.18b]$$

and the unavailability $Q(t)$ of a module is subsequently determined by:

$$Q(t) = (1 - P(t)) = 1 - p^M(t). \qquad [3.18c]$$

The equivalent failure rate of a module $\lambda_m(t)$ determined above by equation [3.2] becomes equal to:

$$\lambda_m(t) = \sum_{m=1}^{M} \lambda(t) = M\lambda(t). \qquad [3.19]$$

The equivalent repair rate of a line $\mu_m(t)$ determined above by equation [3.3] becomes equal to:

$$\mu_m(t) = \sum_{j=1}^{M} \left( \frac{\mu(t)q(t)}{Q(t)} \prod_{m=1}^{M-1} p(t) \right) = \frac{M\mu(t)q(t)p^{M-1}(t)}{Q(t)}. \qquad [3.20]$$

From now on, the $N$ modules can be represented as elementary and binary components, because of equations [3.13–3.15].

The probability $W_l(t)$ that the system is in the set $\mathcal{L}_l^N$ determined above using equation [3.5] becomes equal to:

$$W_l(t) = \sum_{k=1}^{k_l} \left( \prod_{j_k=1}^{l} Q(t) \cdot \prod_{i_k=1}^{N-l} P(t) \right) = \binom{N}{l} Q^l(t) \cdot P^{N-l}(t). \qquad [3.21]$$

We can then determine the equivalent rate of degradation $\lambda_l(t)$ corresponding to the set $\mathcal{L}_l^N$ using equation [3.13] that will be simplified as following:

$$\Lambda_l(t) \cdot W_l(t) = \sum_{k=1}^{k_l} \left[ \left( \sum_{i_k=1}^{N-l} \lambda_{i_k}(t) \right) \left( \prod_{i_k=1}^{N-l} P_{i_k}(t) \right) \left( \prod_{j_k=1}^{l} Q_{j_k}(t) \right) \right]$$
$$= \binom{N}{l} (N-l)\lambda_m(t) \cdot Q^l(t) \cdot P^{(N-l)}(t) = \binom{N}{l} (N-l)\lambda_m(t) \cdot W_l(t).$$

Subsequently, $\Lambda_l(t)$ will be determined by:

$$\Lambda_l(t) = \binom{N}{l}(N-l)\lambda_m(t), \qquad [3.22]$$

where $\lambda_m$ is determined by equation [3.19].

The equivalent rate of recovery $\mu_l(t)$ corresponding to the set $\mathcal{L}_l^N$ and given in equation [3.14] will be determined by:

$$M_l(t) \cdot W_l(t) = \sum_{k=1}^{k_l}\left[\left(\sum_{j_k=1}^{l}\mu_{j_k}\right)\left(\prod_{i_k=1}^{N-l}P_{j_k}(t)\right)\left(\prod_{j_k=1}^{l}Q_{j_k}(t)\right)\right]$$
$$= \binom{N}{l} l\, \mu_m(t) \cdot Q^l(t) \cdot P^{(N-l)}(t) = \binom{N}{l} l\, \mu_m(t) \cdot W_l(t).$$

Thus:

$$\mathrm{M}_l(t) = \binom{N}{l} l\, \mu_m(t), \qquad [3.23]$$

where $\mu_m$ is determined by equation [3.20].

## 3.4. Academic case

A group of thermal power stations for the production of electricity is supplied by five independent natural gas supply lines with identical capacity and functional characteristics.

### 3.4.1. *Case specifications*

Each line consists of five identical pumping stations well defined by their characteristic operating parameters $(\lambda, \mu, \gamma)$ as follows:

$$\lambda = 2{,}0 \times 10^{-2} d^{-1}, \qquad \mu = 2{,}5 \times 10^{-1} d^{-1}, \qquad \gamma = 0.$$

The gas supply lines are designed to transport, under safe conditions, up to 133% of their nominal transport load in case of need.

We are dealing with a $5 \times 5$ matrix system.

### 3.4.2. *Determine system performance*

The system performance assessment will evaluate the loss in electrical production of the group due to potential degradation in the natural gas supply.

### 3.4.3. *Results*

For each gas transmission line, the equivalent characteristic parameters of the module representing the five pumping stations are determined by applying equations [3.18–3.20] and using the characteristic parameters of the stations ($\lambda = 2.0 \times 10^{-2} d^{-1}, \mu = 2.5 \times 10^{-1} d^{-1}, \gamma = 0$). Subsequently, we have five identical modules.

The five sets $\mathcal{L}_l^N$, ($\mathcal{L}_l^N; l = 1,2,3,4,5$) representing the five possible degradation phases of the gas supply system are identified. Then, we determine the probabilities of being in each operating phase $W_l(t)$ $\{W_l(t); l = 0,1,2,3,4,5\}$.

The percentage of gas supply loss $\epsilon_l$ associated with each phase of degraded operation $l$ is calculated (Table 3.3). It is noted that no gas supply losses are expected when only one module is unavailable, that is, in the degraded state $W_1(t)$.

The expected gas supply loss percentage associated with each degraded phase is weighted by its probability (Figure 3.8):

$$\delta_l(t) = \epsilon_l W_l(t), \qquad l = 2, 3, 4, 5.$$

The expected overall gas supply loss percentage $\hat{\delta}(t)$ as a function of time is estimated and traced in Figure 3.9:

$$\hat{\delta}(t) = \sum_{l=1}^{5} \delta_l(t).$$

## 3.5. Conclusion

It is obvious that systems performance analysis is not linked only to the nature of a system to be analyzed: whether it is a matrix-like system or not, homogeneous or not. Our choice to assess the performance analysis of matrix-like systems is an arbitrary choice. It results from our experience in systems engineering where most of the large matrix systems are related to well-defined production or supplying processes. In this context, performance analysis reveals all its analytical power. The performance analysis is beyond the reliability-availability or risk assessments even if it uses the same probabilistic basic knowledge and skills.

In this context, performance assessments do not relate to the evaluation of unavailability only, but relates ultimately to the evaluation of the expected productivity or degradation of productivity of a given system.

| $N^b$ lines indi. | Increase required/line avail. in % | Increase granted/ line avail. in % | Supply loss $\epsilon_l$ in % | Prob. |
|---|---|---|---|---|
| 0 | $\frac{0}{5-0} = 0 \leq 33$ | 0 | $(100 - 0 \times 0) = 0$ | $W_0(t)$ |
| 1 | $\frac{1}{5-1} = 25 \leq 33$ | 25 | $(100 - 4 \times 25) = 0$ | $W_1(t)$ |
| 2 | $\frac{2}{5-2} = 60 > 33$ | 33 | $(100 - 3 \times 33) = 1$ | $W_2(t)$ |
| 3 | $\frac{3}{5-3} = 150 > 33$ | 33 | $(100 - 2 \times 33) = 34$ | $W_3(t)$ |
| 4 | $\frac{4}{5-4} = 400 > 33$ | 33 | $(100 - 1 \times 33) = 67$ | $W_4(t)$ |
| 5 | $\frac{5}{5-5} = \infty > 33$ | 33 | $(100 - 0 \times 20) = 100$ | $W_5(t)$ |

**Table 3.3.** *The percentage of loss of gas supply*

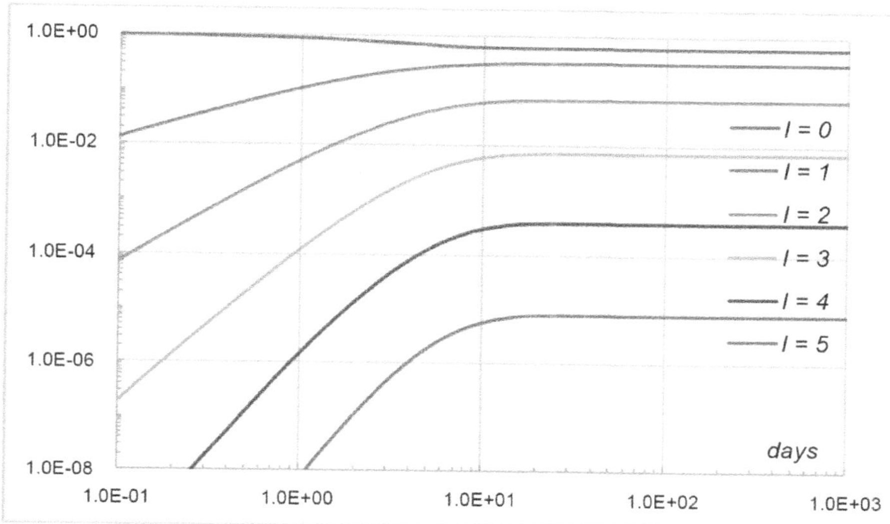


**Figure 3.7.** *The probability of being in a degraded phase $W_l(t)$ depending on the number of unavailable modules ($l = 0,1,2, \ldots ,5$). For a color version of this figure, see www.iste.co.uk/elhami/reliability.zip*

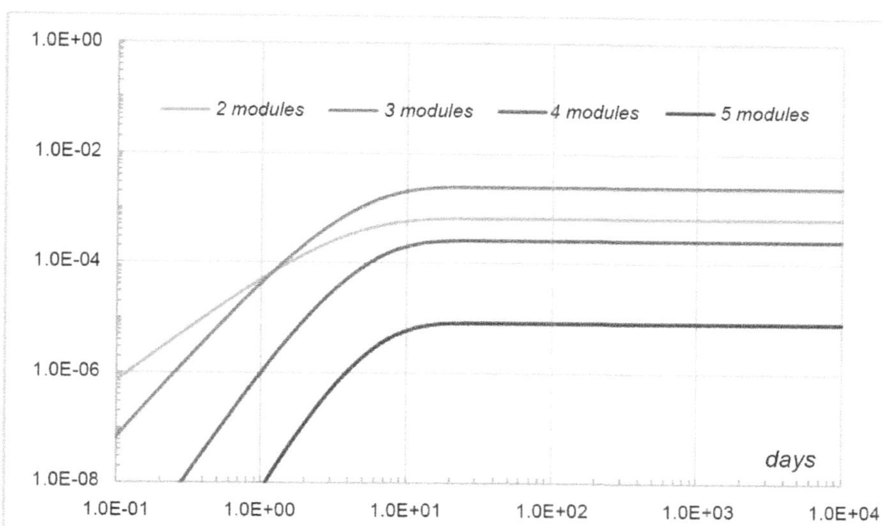


**Figure 3.8.** *Expected supply loss percentage per degradation phase $\mathcal{L}_l^N$ ($\mathcal{L}_l^N; l = 0,1,2,3,4,5$). For a color version of this figure, see www.iste.co.uk/elhami/reliability.zip*

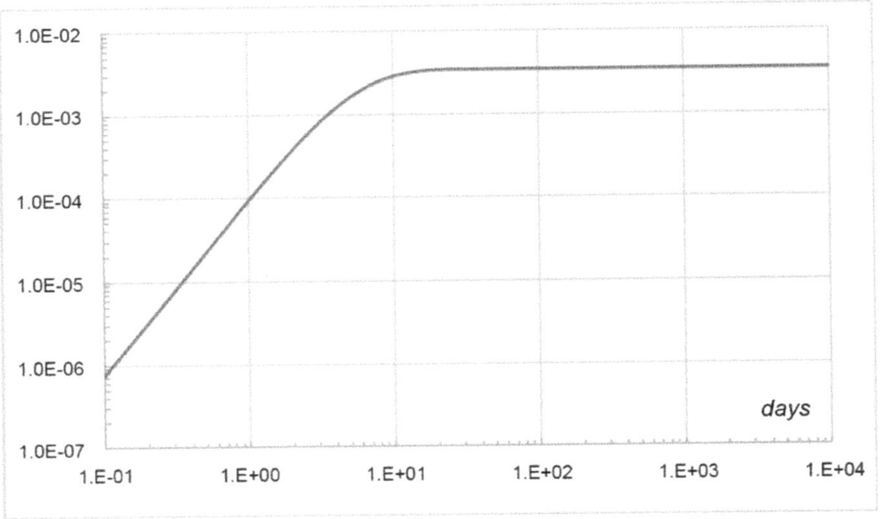


**Figure 3.9.** *Overall expected supply loss percentage. For a color version of this figure, see www.iste.co.uk/elhami/reliability.zip*

# 4

# Modeling of Systems with Redundancy

## 4.1. Introduction

Redundancy is a technical solution that generally improves the availability and the reliability of systems at higher total cost for installation and maintenance. So, we are talking about the system with typical ${}^{n}/_{N}$ redundancy. In the literature, there are two definitions of the ${}^{n}/_{N}$ system. First, it is a macrosystem made up of $N$ systems, functionally identical, and is available if at least $n$ of its systems are available. Second, it is a macrosystem made up of $N$ systems, functionally identical, and is unavailable if at least $n$ of its systems are unavailable. In the first definition, $n$ indicates the minimal availability condition of the macrosystem. In the second definition, $n$ indicates the minimal unavailability condition of the macrosystem. In this chapter, we opt for the first definition. The functional resemblance between the $N$ systems does not necessarily mean that these systems are identical physically, structurally, technologically or in operational performance. Systems with redundancy represent a subset of matrix-like systems covered in Chapter 3. The fundamental distinction results from the existence of a minimal condition of unavailability (or availability).

It is worth noting that the functional analysis of matrix systems (Chapter 3) essentially focuses on the classification of the functional states of the system according to the number of the unavailable (or available) modules/components. It is a classification according to the state of degradation of the system. This is natural in the context of performance evaluation.

As for the analysis of the unavailability and the reliability of the macrosystem, the classification of the states is based on a single criterion that is its unavailability (or its availability), often decreed by minimal conditions. From this perspective, the functional space of the macrosystem is split into two disjoint spaces: the first contains a set of states where the macrosystem is available and the second contains a set of states where the macrosystem is unavailable. We are back to a binary logic reasoning scheme with two disjoint spaces separated by a threshold, determined by a minimal condition.

The improvement expected by the redundancy requires that the constituent systems are effectively independent: structurally and functionally. Otherwise, the dependency between the constituent systems, whatever its origin, systematically gives rise to “common mode failures”, with the risk of degrading the availability and the reliability of the macrosystem. Thus, redundancy loses its interest.

Throughout this chapter, we deal with active redundancy. Passive redundancy requires considering the startup of the systems that are on standby. This would have weighed down the models used here and masked the didactic purpose of the chapter. However, extending the model to passive redundancy cases is straight forward.

## 4.2. Combinatorial analysis of the *n/N* system

A system $S_R$ with redundancy $n/N$ indicates that it consists of $N$ systems $S_i$, $(i = 1,2, \dots, N)$ and that we need at least $n$ available systems to carry out a specific mission. If $k_n$ is the number of possible combinations resulting from grouping the $S_i$ systems into groups of size $n$, $k_n$ is determined by:

$$k_n = \binom{N}{n} = \frac{N!}{n!(N-n)!}. \tag{4.1}$$

Consider an $n/N$ system that is available if at least $n$ of the $N$ constituent systems are available, then the total number of combinations $k_{\geq n}$ included in the availability space will be determined by:

$$k_{\geq n} = \sum_{i=n}^{N} k_i, \tag{4.2a}$$

and the total number of combinations $k_{<n}$ included in the space of unavailability will be determined by:

$$k_{<n} = \sum_{i=0}^{n-1} k_i. \qquad [4.2b]$$

Note that:

$$k_{<n} + k_{\geq n} = 2^N. \qquad [4.3]$$

Now we know the number $k_{\geq n}$ of states that make the system available. They are all identified and indexed by $i(\ i \in [n, N])$. In addition, we also know the number $k_{<n}$of states that make the system unavailable. They are each identified and indexed by $i(i \in [0, (n-1)])$. We can therefore determine the unavailability and the availability of an *n*/*N* redundant system.

## 4.3. Determining the availability/unavailability

To determine the availability and/or the unavailability of a system with *n*/*N* redundancy, one can use two approaches: the analysis by the binary structure function (Chapter 3) and the Boolean algebraic approach. We can also use the state graph analysis as already covered in Chapter 2.

Given that the availability of the system $S_{n/N}$ is defined by a minimal condition, we can easily describe the states of the redundant system by binary structure functions or by Boolean expressions.

### 4.3.1. *Analysis by the binary structure function*

We identify the sets $\mathcal{L}_l^N$, $l\ (l = 0,1,2, \dots, N)$, included in the functional space of the system, following the approach previously presented in Chapter 3. Then, we determine the probabilities $W_l(t)$ of being in each set by equation [3.5] (see details in Chapter 3).

Since the minimal condition of availability is defined by $n$ available systems among $N$, the availability of the system $P_{\text{sys}}(t)$ will be determined by:

$$P_{\text{sys}}(t) = \sum_{l=0}^{N-n} W_l(t), \qquad [4.4a]$$

and system $Q_{\text{sys}}(t)$ unavailability will be determined by:

$$Q_{\text{sys}}(t) = \sum_{l=N-n+1}^{N} W_l(t), \qquad [4.4b]$$

where $l$ indicates the number of the unavailable constituent systems (see the definition of $W_l(t)$ in Chapter 3). We have already treated a numerical application by using the previous formalisms in Chapter 3. Mind that the index $l$, in equations [4.4], describes the number of the unavailable constituent systems as the focus in Chapter 3 was on the degradation of the matrix-like system.

We will rather use here the Boolean expressions in modeling the unavailability, the availability and the characteristic parameters $(\lambda, \mu, \gamma)$ of the redundant system $S_{n/N}$. Thus, we are interested in the binarization of the redundant system.

### 4.3.2. *Analysis by Boolean expressions*

The minimal condition of the availability of the macrosystem is described by the availability of at least $n$ systems among $N$. By complementarity, the minimal condition of the unavailability of the macrosystem will be described by the unavailability of at least $(N - n + 1)$ systems among $N$. Then, the availability space must be defined by $\binom{N}{n}$ minimal cuts and the unavailability space by $\binom{N}{N-n+1}$ minimal cuts.

Take the case of a $^2/_4$ redundancy, which means that only two available systems out of four are needed so that the macrosystem may be available. But this also means that three out of four unavailable systems would suffice to switch the macrosystem to the unavailability state. The macrosystem would have six minimal cuts $\left[\binom{4}{2} = 6\right]$ of order 2 in its Boolean expression of availability. But it would have four minimal cuts $\left[\binom{4}{3} = 4\right]$ of order 3 in its Boolean expression of unavailability (see details in section 4.3.3). An analyst concerned with the economy of effort would probably choose to proceed with the expression of system unavailability in this case.

So, the space of availability $\Omega$ will generally be described by the following Boolean expression:

$$\begin{aligned}\Omega = {} & S_1 \cdot S_2 \cdot \ldots \cdot S_n + S_1 \cdot S_3 \cdot \ldots \cdot S_{n+1} + S_1 \cdot S_4 \cdot \ldots \cdot S_{n+2} + \cdots \\ & + S_2 \cdot S_3 \cdot \ldots \cdot S_{n+1} + S_2 \cdot S_4 \cdot \ldots \cdot S_{n+2} + S_1 \cdot S_5 \cdot \ldots \cdot S_{n+3} + \cdots \\ & + \ldots\ldots\ldots\ldots\ldots\ldots\ldots\ldots\ldots\ldots\ldots\ldots\ldots\ldots\ldots\ldots\ldots \\ & + S_{N-n+1} \cdot S_{N-n+2} \cdot \ldots \cdot S_{N-2} \cdot S_{N-1} \cdot S_N. \end{aligned} \quad [4.5]$$

We can also describe the unavailability space $\overline{\Omega}$ of the system by:

$$\begin{aligned}\overline{\Omega} = {} & \overline{S_1} \cdot \overline{S_2} \cdot \ldots \cdot \overline{S_{N-n+1}} + \overline{S_1} \cdot \overline{S_3} \cdot \ldots \cdot \overline{S_{N-n+2}} + \overline{S_1} \cdot \overline{S_4} \cdot \ldots \cdot \overline{S_{N-n+3}} + \cdots \\ & + \overline{S_2} \cdot \overline{S_3} \cdot \ldots \cdot \overline{S_{N-n+2}} + \overline{S_2} \cdot \overline{S_4} \cdot \ldots \cdot \overline{S_{N-n+3}} + \overline{S_2} \cdot \overline{S_5} \cdot \ldots \cdot \overline{S_{N-n+4}} + \cdots \\ & + \ldots\ldots\ldots\ldots\ldots\ldots\ldots\ldots\ldots\ldots\ldots\ldots\ldots\ldots\ldots\ldots\ldots \\ & + \overline{S_n} \cdot \overline{S_{n+1}} \cdot \overline{S_{n+2}} \cdot \ldots \cdot \overline{S_{N-2}} \cdot \overline{S_{N-1}} \cdot \overline{S_N}, \end{aligned} \quad [4.6]$$

where $n$ is the minimal number of available systems. $S_i$ and $\overline{S_i}$ are two Boolean functions describing the availability and the unavailability of the $i$th constituent system ($i = 1, 2, \ldots, N$), respectively. The sign "+" designates the Boolean operator "OR" and the sign "·" designates the operator "AND". Equations [4.5] and [4.6] are two logical expressions declaring the same functional reality from two different complementary points of view: the availability and unavailability of redundant systems.

Equation [4.5] contains the set of minimal cuts necessary to make the macrosystem available. The expression in equation [4.6] contains the set of minimal cuts that make the macrosystem unavailable.

Note that:

$$\Omega + \overline{\Omega} = U,$$

$$\Omega \cdot \overline{\Omega} = \emptyset.$$

The spaces $\Omega$ and $\overline{\Omega}$ are complementary and disjoint. Their universal space $U$ consists of $2^N$ states. Spaces $\Omega$ and $\overline{\Omega}$ do not share any common state ($\emptyset$).

The availability of the macrosystem $P_{\text{sys}}(t)$ is determined by the probability $P(\Omega)$ to be in the space $\Omega$, and the unavailability of the macrosystem $Q_{sys}(t)$ is determined by the probability $Q(\overline{\Omega})$ of being in the space $\overline{\Omega}$. It is obvious that it suffices to determine one of these two probabilities, given that:

$$P(\Omega) + Q(\overline{\Omega}) = 1.$$

Expressions with minimal cuts are very useful because they directly express the availability and unavailability in declarative clear terms to analysts and can be used directly in the qualitative analysis of systems. However, they cannot be used directly with their shapes in equations [4.5 and 4.6] to easily quantify $P(\Omega)$ and $Q(\overline{\Omega})$. It is necessary to rearrange these logical expressions so that they become directly usable by numerical applications.

One of the ways to make them directly usable is to transform the minimal cuts into disjoint cuts, because of the basic rules of Boolean algebra, as follows.

Consider a set of states $\Omega$ defined by subsets belonging to it, as follows:

$$\Omega = \Omega_1 + \Omega_2 + \Omega_3 + \cdots + \Omega_N \qquad [4.7]$$

where:

$$\Omega_i \subset \Omega, \forall\, i = 1,2, \ldots, N.$$

We can rewrite the expression [4.7] in an equivalent form:

$$\begin{aligned}\Omega &= \Omega_1 + \overline{\Omega_1} \cdot \Omega_2 + \overline{\Omega_1} \cdot \overline{\Omega_2} \cdot \Omega_3 + \cdots + \overline{\Omega_1} \cdot \overline{\Omega_2} \cdot \ldots \cdot \overline{\Omega_{N-1}} \cdot \Omega_N \\ &= \Omega_1^* + \Omega_2^* + \Omega_3^* + \cdots + \Omega_N^*,\end{aligned} \qquad [4.8]$$

where:

$$\Omega_i^* \subset \Omega,\ \Omega_i^* \cdot \Omega_{j \neq i}^* = \emptyset,\ \forall\, i, j = 1,2, \ldots, N.$$

The subsets $\Omega_i$ and $\overline{\Omega_i}$ are complementary subsets. These subsets can be any cuts: minimal or not and independent or not.

The interest of expression [4.8] lies in the fact that the new cuts $\Omega_i^*$ are mutually disjoint. Obviously, the cuts $\Omega_i^*$ are no longer minimal. The Boolean expression in equation [4.8] is equivalent to the expression in equation [4.7]. There are $[(k_n)!]$ Boolean expressions equivalent to the expression in equation [4.7], where $k_n$is determined by:

$$k_n = \binom{N}{n}.$$

If one can separately determine the probabilities $P(\Omega_i^*)$, the probability $P(\Omega)$of being in space $\Omega$ can be determined directly by the following sum:

$$P_{\text{sys}}(t) = P(\Omega) = \sum_{i=1}^{N} P(\Omega_i^*). \quad [4.9a]$$

Similarly, for unavailability:

$$Q_{\text{sys}}(t) = Q(\overline{\Omega}) = \sum_{i=1}^{N} Q(\overline{\Omega_i^*}). \quad [4.9b]$$

Remember that it suffices to determine either the availability $P(\Omega)$ or the unavailability $Q(\overline{\Omega})$.

In the following, we are going to propose an exercise to illustrate the approach of the analysis using Boolean expressions.

### 4.3.3. *Example of analysis using Boolean expressions*

Consider $S_{2/4}$ a system with redundancy $2/4$ consisting of four identical constituent systems $S_i$. The system $S_{2/4}$ is available if and only if at least two systems $S_i$ are available. So, we will have six minimal cuts $\left(\binom{4}{2} = 6\right)$. The space of availability $\Omega$ will be described by the following Boolean expression:

$$\Omega = S_1 \cdot S_2 + S_1 \cdot S_3 + S_1 \cdot S_4 + S_2 \cdot S_3 + S_2 \cdot S_4 + S_3 \cdot S_4. \quad [4.10]$$

Equation [4.10] describes the availability space $\Omega$ by the minimal cuts. However, the availability space $\Omega$ can also be described by the equivalent disjoint cuts as in the following Boolean expression:

$$\Omega = S_1 \cdot S_2 + \overline{(S_1 \cdot S_2)} \cdot S_1 \cdot S_3 + \overline{(S_1 \cdot S_2)} \cdot \overline{(S_1 \cdot S_3)} \cdot S_1 \cdot S_4$$
$$+ \overline{(S_1 \cdot S_2)} \cdot \overline{(S_1 \cdot S_3)} \cdot \overline{(S_1 \cdot S_4)} \cdot S_2 \cdot S_3$$
$$+ \overline{(S_1 \cdot S_2)} \cdot \overline{(S_1 \cdot S_3)} \cdot \overline{(S_1 \cdot S_4)} \cdot \overline{(S_2 \cdot S_3)} \cdot S_2 \cdot S_4$$
$$+ \overline{(S_1 \cdot S_2)} \cdot \overline{(S_1 \cdot S_3)} \cdot \overline{(S_1 \cdot S_4)} \cdot \overline{(S_2 \cdot S_3)} \cdot \overline{(S_2 \cdot S_4)} \cdot S_3 \cdot S_4$$

$$\Omega = S_1 \cdot S_2 + S_1 \cdot \overline{S_2} \cdot S_3 + S_1 \cdot \overline{S_2} \cdot \overline{S_3} \cdot S_4 + \overline{S_1} \cdot S_2 \cdot S_3 + \overline{S_1} \cdot S_2 \cdot \overline{S_3} \cdot S_4 + \overline{S_1} \cdot \overline{S_2} \cdot S_3 \cdot S_4. \quad [4.11]$$

Because of the Boolean expression in equation [4.11] using the disjoint cuts, we can also determine the number of states $Nb(\Omega)$ belonging to the availability space $\Omega$, as follows:

$$Nb(\Omega) = 2^{4-2} + 2^{4-3} + 2^{4-4} + 2^{4-3} + 2^{4-4} + 2^{4-4} = 4 + 2 + 1 + 2 + 1 + 1 = 11.$$

Alternatively, we can define the space of unavailability of the system $S_{2/4}$ which is unavailable if and only if at least three systems $S_i$ are unavailable. We will therefore have four minimal cuts $\left(\binom{4}{3} = 4\right)$. The space of unavailability $\overline{\Omega}$ is described by the following Boolean expression:

$$\overline{\Omega} = \overline{S_1} \cdot \overline{S_2} \cdot \overline{S_3} + \overline{S_1} \cdot \overline{S_2} \cdot \overline{S_4} + \overline{S_1} \cdot \overline{S_3} \cdot \overline{S_4} + \overline{S_2} \cdot \overline{S_3} \cdot \overline{S_4}. \quad [4.12]$$

Equation [4.12] describes the space of unavailability $\overline{\Omega}$ by the minimal cuts. The space of unavailability $\overline{\Omega}$ can also be described by the disjoint cuts as in the following Boolean expression:

$$\overline{\Omega} = \overline{S_1} \cdot \overline{S_2} \cdot \overline{S_3} + \overline{(\overline{S_1} \cdot \overline{S_2} \cdot \overline{S_3})} \cdot \overline{S_1} \cdot \overline{S_2} \cdot \overline{S_4}$$
$$+ \overline{(\overline{S_1} \cdot \overline{S_2} \cdot \overline{S_3})} \cdot \overline{(\overline{S_1} \cdot \overline{S_2} \cdot \overline{S_4})} \cdot \overline{S_1} \cdot \overline{S_3} \cdot \overline{S_4}$$
$$+ \overline{(\overline{S_1} \cdot \overline{S_2} \cdot \overline{S_3})} \cdot \overline{(\overline{S_1} \cdot \overline{S_2} \cdot \overline{S_4})} \cdot \overline{(\overline{S_1} \cdot \overline{S_3} \cdot \overline{S_4})} \cdot \overline{S_2} \cdot \overline{S_3} \cdot \overline{S_4}.$$

After reduction, we obtain the expression of the unavailability with these disjoint cuts, as follows:

$$\overline{\Omega} = \overline{S_1} \cdot \overline{S_2} \cdot \overline{S_3} + \overline{S_1} \cdot \overline{S_2} \cdot S_3 \cdot \overline{S_4} + \overline{S_1} \cdot S_2 \cdot \overline{S_3} \cdot \overline{S_4} + S_1 \cdot \overline{S_2} \cdot \overline{S_3} \cdot \overline{S_4}. \quad [4.13]$$

The number of states $Nb(\overline{\Omega})$, belonging to the unavailability space $\overline{\Omega}$, is determined as follows:

$$Nb(\overline{\Omega}) = 2^{4-3} + 2^{4-4} + 2^{4-4} + 2^{4-4} = 2 + 1 + 1 + 1 = 5.$$

We already know that the universal functional space of a four-component system must have $2^4$ total states. Thus:

$$Nb(\Omega) + Nb(\overline{\Omega}) = 16 = 2^4.$$

Whichever method is used, "structure binary function" or "Boolean logical expression", we can determine the availability $P(\Omega)$ and the unavailability $Q(\overline{\Omega})$ of the redundant system. We therefore have to determine the equivalent failure and repair rates to complete the characterization of the system.

## 4.4. The equivalent failure and repair rates

In order to determine the equivalent transition rates, it is necessary to identify and isolate the critical states. Maintaining the assumption of independence between the constituent systems, the set of critical availability states or the set of critical unavailability states are, respectively, those which contain exactly $n$ available systems or $(N - n + 1)$ unavailable systems, except $N$.

Let us focus on determining the equivalent failure rate. The set of critical availability states contains exactly $K$ states determined by:

$$K = \binom{N}{n}.$$

We identify these states and renumber them with the index $k$ from one to $K$. In every state $k$, the available systems are renumbered with the index $i_k$ from one to $n$ (see section 3.2.2). By following the procedure of section 3.2.5, we can determine the equivalent failure rate $\Lambda_{sys}(t)$ by:

$$\Lambda_{sys}(t)P_{sys}(t) = \sum_{k=1}^{K}\left(\sum_{i_k=1}^{n}\lambda_{i_k}P_{i_k}(t)\cdot\prod_{\substack{j_k=1\\ j_k\neq i_k}}^{N}Q_{j_k}(t)\right) \qquad [4.14]$$

where $P_{i_k}(t)$ and $Q_{j_k}(t)$ are, respectively, the availabilities for the $n$ available systems ( $i_k$; $i_k = 1, 2, \dots, n$) and the unavailabilities for the remaining systems ($j_k$; $j_k = 1, 2, \dots, N$; $j_k \neq i_k$), and $P_{sys}(t)$ is the availability of the system with redundancy $^n/_N$.

Similarly, for the set of critical unavailability states, there are exactly $K$ states determined by:

$$K = \binom{N}{(N-n+1)}.$$

These states are identified and renumbered with the index $k$ ranging from one to $K$. In every state $k$, unavailable systems are renumbered with the index $j_k$ from one to $(N - l + 1)$ (see section 3.2.2). By following the procedure in section 3.2.5, we can also determine the equivalent repair rate $M_{sys}(t)$ by:

$$M_{sys}(t) Q_{sys}(t) = \sum_{k=1}^{K} \left( \sum_{j_k=1}^{N-l+1} \mu_{j_k} Q_{j_k}(t) \cdot \prod_{\substack{i_k=1 \\ i_k \neq j_k}}^{N} P_{i_k}(t) \right), \quad [4.15]$$

where $Q_{j_k}(t)$ and $P_{i_k}(t)$ are, respectively, the unavailabilities of the $(N-n+1)$ unavailable systems ($j_k$; $j_k = 1, 2, \dots, N-n+1$) and the availabilities of the remaining available systems ($i_k$; $i_k = 1, 2, \dots, N$; $i_k \neq j_k$), and $Q_{sys}(t)$ is the unavailability of the system with redundancy $^n/_N$.

As for the initial condition $\gamma_{sys}$ of the system with redundancy $^n/_N$, it is determined by:

$$\Gamma_{sys} = Q_{sys}(0). \quad [4.16]$$

The characteristic parameters $(\Lambda(t), M(t), \Gamma(t))_{sys}$ of the system with redundancy $^n/_N$ are now determined.

If all the systems constituting the macrosystem $S_{n/N}$ are operationally identical, that is, if they all have the same characteristic parameters $(\lambda(t), \mu(t), \gamma(t))$, the macrosystem will be described as homogeneous.

## 4.5. Homogeneous system

If, and only if, all constituent systems, $S_i$ $(i = 1, 2, \ldots, N)$, belonging to the redundant system $S_{n/N}$ are operationally identical, these systems will be characterized by the same availability $p(t)$, the same unavailability $q(t)$ and the same characteristic parameters $(\lambda(t), \mu(t), \gamma)$.

The homogeneity of the constituents of the redundant system ${}^{n}/_{N}$ significantly simplifies the mathematical expressions of the availability $P_{\text{sys}}(t)$, the unavailability $Q_{\text{sys}}(t)$ and the characteristic parameters $(\Lambda_{\text{sys}}(t), M_{\text{sys}}(t), \Gamma_{\text{sys}})$.

### 4.5.1. *Characterization of the homogeneous system*

Considering equations [4.4a] and [4.4b], the availability $P_{\text{sys}}(t)$ and unavailability $Q_{\text{sys}}(t)$ of the redundant system are determined by:

$$P_{\text{sys}}(t) = \sum_{j=n}^{N} \binom{N}{j} \times (p(t))^{j} \times (q(t))^{N-j}, \qquad [4.17]$$

$$Q_{\text{sys}}(t) = \sum_{j=0}^{n-1} \binom{N}{j} \times (p(t))^{j} \times (q(t))^{N-j}, \qquad [4.18]$$

where $j$ represents the number of the available constituent systems and $n$ is the minimal availability condition.

In this case, the equivalent failure rate $\Lambda_{\text{sys}}(t)$ of the redundant system is determined by:

$$\Lambda_{sys}(t) P_{sys}(t) = \binom{N}{n} \times n\lambda \times (p(t))^{n} \times (q(t))^{N-n}. \qquad [4.19]$$

The equivalent repair rate $M_{\text{sys}}(t)$ is determined by:

$$M_{sys}(t) Q_{sys}(t) = \binom{N}{N-n+1} \times (N-n+1) \times \mu \times (p(t))^{n-1} \times (q(t))^{N-n+1}, \qquad [4.20]$$

and, the initial condition $\Gamma_{sys}$ will be determined by:

$$\Gamma_{sys} = \sum_{j=0}^{n-1} \binom{N}{j} \times (1-\gamma)^j \times (\gamma)^{N-j}, \qquad [4.21]$$

where $n$ is the minimal number of available systems necessary for the macrosystem.

## 4.6. Numerical application (1)

Consider $S_i$ a system consisting of two elementary components $C1$ and $C2$ that operate in parallel. The system fails if and only if both components fail. The characteristic parameters $(\lambda, \mu, \gamma)$ of the components are $C1(10^{-4}\text{h}^{-1}, 10^{0}\text{h}^{-1}, 3\times10^{-1})$ and $C2(10^{-2}\text{h}^{-1}, 2\times10^{-2}\text{h}^{-1}, 0)$. This system has already been treated in Chapter 2. Its unavailability and its equivalent failure rate have already been determined in Figures 2.2 and 2.4.

A well-defined mission could be carried out with, at least, two identical units of the type $S_i$, placed in parallel. To improve the system availability and reliability, a third unit $S_i$ is added to create a 2/3 redundant system $S_{2/3}$.

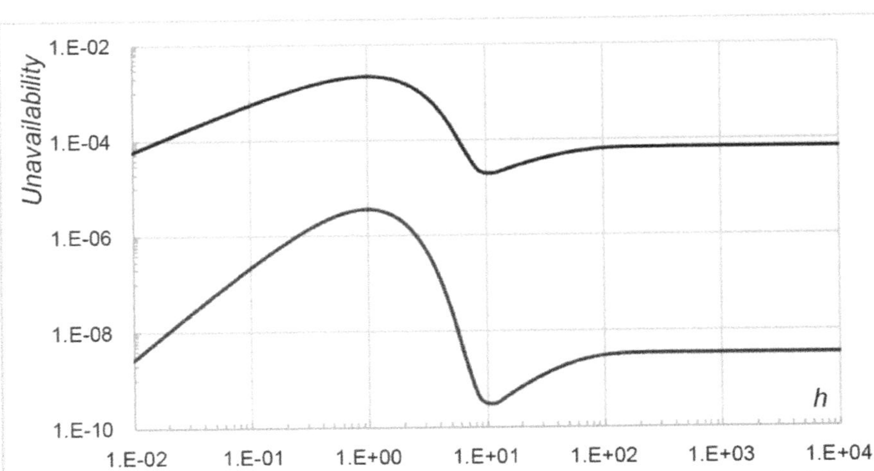


**Figure 4.1.** *Comparison between the unavailability without redundancy (in black) and unavailability with 2/3 redundancy (in red). For a color version of this figure, see www.iste.co.uk/elhami/reliability.zip*

To characterize this system with 2/3 redundancy, we use the simplified formulas expressed by equations [4.7–4.21], using as input data the results obtained in Chapter 3 for the system $S_i$, $i = 1,2,3$.

Regarding the unavailability, we compare in Figure 4.1 the systems with two units of type $S_i$ (without redundancy) with a 2/3 redundancy system. We find that the 2/3 redundancy allowed us to reduce the unavailability by more than four decades over the interval from startup to the asymptote, $t \in [0, \infty]$.

Regarding the equivalent failure rate, we compare in Figure 4.2 the two situations: without redundancy and with 2/3 redundancy. We find that the 2/3 redundancy allowed us to reduce the equivalent failure rate by more than four decades over the same time interval $t$ ($t \in [0, \infty[$). We also identify three distinct phases on the failure rate profiles.

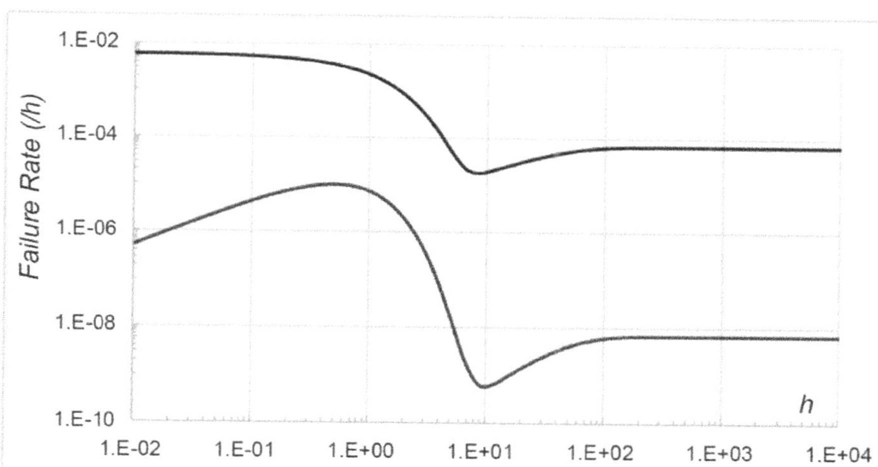


**Figure 4.2.** *Comparison between the equivalent failure rate without redundancy (in black) and with 2/3 redundancy (in red). For a color version of this figure, see www.iste.co.uk/elhami/reliability.zip*

These phases are as follows: aging between 0 h and 1 h which is of functional origin due to startup difficulties ($\gamma \neq 0$), rejuvenation between 1 h and 10 h (healing from the startup effect) and aging between 10 h and 100 h before attending an asymptotic value lower than the values at startup. The system (with or without redundancy) suffers from a problem at startup.

Redundancy relatively amplifies this problem during the first hour of operation.

## 4.7. Numerical application (2)

This is a fire detection system, in a carpentry workshop, made up of three independent detectors $(L1, L2, L3)$. The three detectors are based on three different technologies and manufactured by three different manufacturers. This is mainly to avoid common mode failures.

The three detectors are installed in the same workshop, but at three different positions. Each detector is characterized by a "mean time before detect (MTBD) – $\tau$" depending on the intrinsic characteristics of the detector and its position in the workshop. The detection system is supposed to detect fire anywhere in the workshop hall.

### 4.7.1. *Mission success criteria and system data*

Fire detection is confirmed if at least two detectors out of three detect it.

The detectors are presumed to be elementary components independent of each other and their basic parameters $(\lambda, \mu, \gamma)$ and their MTBD " $\tau$" are given in Table 4.1.

| # | Making up | $\lambda(\mathrm{h}^{-1})$ | $\mu(\mathrm{h}^{-1})$ | $\gamma$ | $\tau(\mathrm{s})$ |
|---|---|---|---|---|---|
| 1 | Detector 1 $(L1)$ | $8.0^{-3}$ | $1.0^{-1}$ | 0 | 10 |
| 2 | Detector 2 $(L2)$ | $4.0^{-3}$ | $8.0^{-1}$ | 0 | 20 |
| 3 | Detector 3 $(L3)$ | $2.0^{-3}$ | $4.0^{-1}$ | 0 | 30 |

**Table 4.1.** *The characteristic parameters of the detectors*

### 4.7.2. *Functional modeling by state graph*

The functional space of the detection system contains eight states $(2^3 = 8)$. The identification of availability and unavailability states will be achieved through the application of the success criteria defined in section 4.7.1. The totality of the states constituting the functional space and their division between the space of availability ($\Omega$) and the space of unavailability ($\overline{\Omega}$) are presented in Figure 4.3.

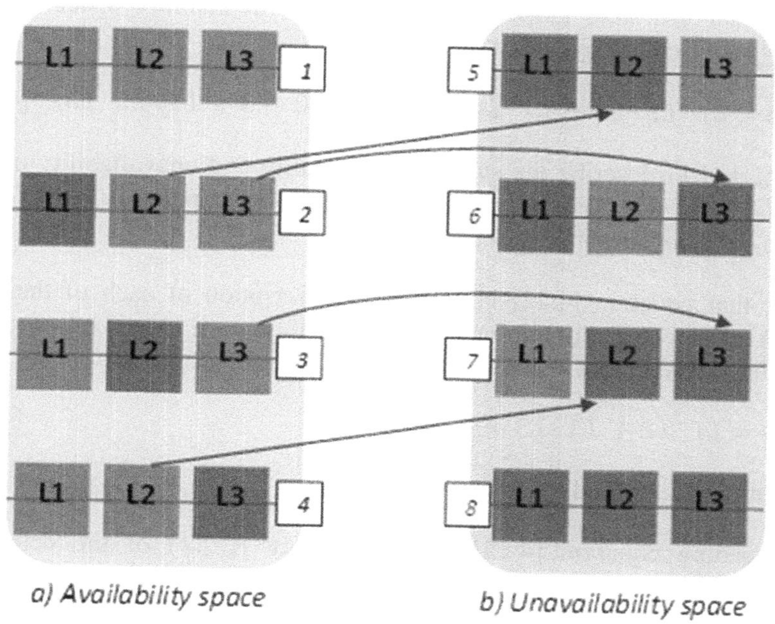


**Figure 4.3.** *(a) The system availability spaces (in green) and (b) the system unavailability spaces (in red). For a color version of this figure, see www.iste.co.uk/elhami/reliability.zip*

We can also identify the corresponding critical transitions. For demonstration purposes, we indicate on the graph certain states that are the sites of critical transitions, such as the state $E2$, and others that experience no critical transition such as $E1$ and $E8$.

### 4.7.3. *System availability modeling*

The availability of the detection system can be modeled directly using the state graph (Figure 4.3). The space of availability $\Omega$ contains four states $(E1, E2, E3, E4)$ and the space of unavailability $\overline{\Omega}$ groups the four other states $(E5, E6, E7, E8)$ (see Figure 4.3). An identification of these states and the corresponding critical transitions are detailed in Tables 4.2 and 4.3, respectively.

Based on the state graph (Figure 4.3), the availability space $\Omega$ can logically be described by:

$$\Omega = L1 \cdot L2 \cdot L3 + \overline{L1} \cdot L2 \cdot L3 + L1 \cdot \overline{L2} \cdot L3 + L1 \cdot L2 \cdot \overline{L3}, \quad [4.22]$$

where $Li$ and $\overline{Li}$ describe the states of availability and unavailability of the detector "$i$". "·" and "+" are, respectively, the logical operators "AND" and "OR" in the Boolean sense of the term.

Note that equation [4.22] is an exact transcription of each of the four states. It does not contain minimal cuts, but disjoint cuts. However, we could always have rewritten it in the form of minimal cuts as follows:

$$\Omega = L1 \cdot L2 + L1 \cdot L3 + L2 \cdot L3. \quad [4.23]$$

Equations [4.22] and [4.23] are equivalent.

By reading equation [4.22], the availability $P_{sys}(t)$ of the detection system can be assessed as follows:

$$\begin{aligned} P_{sys}(t) = {} & p_1(t)p_2(t)p_3(t) + q_1(t)p_2(t)p_3(t) \\ & + p_1(t)q_2(t)p_3(t) + p_1(t)p_2(t)q_3(t), \end{aligned} \quad [4.24]$$

where $p_i(t)$and $q_i(t)$are the availability and unavailability of the detector $i$ (see equation [1.3]).

| State | Logical expression | Probability of being in the state | Critical transition rate | Non-critical transition rate |
|---|---|---|---|---|
| *E*1 | $L1 \cdot L2 \cdot L3$ | $p_1(t)p_2(t)p_3(t)$ | 0 | $\lambda_1 + \lambda_2 + \lambda_3$ |
| *E*2 | $\overline{L1} \cdot L2 \cdot L3$ | $q_1(t)p_2(t)p_3(t)$ | $\lambda_2 + \lambda_3$ | $\mu_1$ |
| *E*3 | $L1 \cdot \overline{L2} \cdot L3$ | $p_1(t)q_2(t)p_3(t)$ | $\lambda_1 + \lambda_3$ | $\mu_2$ |
| *E*4 | $L1 \cdot L2 \cdot \overline{L3}$ | $p_1(t)p_2(t)q_3(t)$ | $\lambda_1 + \lambda_2$ | $\mu_3$ |

**Table 4.2.** *Availability states and their critical transitions*

| State | Logical expression | Probability of being in the state | Critical transition rate | Non-critical transition rate |
|---|---|---|---|---|
| *E5* | $\overline{L1}\cdot\overline{L2}\cdot L3$ | $q_1(t)q_2(t)p_3(t)$ | $\mu_1+\mu_2$ | $\lambda_3$ |
| *E6* | $\overline{L1}\cdot L2\cdot\overline{L3}$ | $q_1(t)p_2(t)q_3(t)$ | $\mu_1+\mu_3$ | $\lambda_2$ |
| *E7* | $L1\cdot\overline{L2}\cdot\overline{L3}$ | $p_1(t)q_2(t)q_3(t)$ | $\mu_2+\mu_3$ | $\lambda_1$ |
| *E8* | $\overline{L1}\cdot\overline{L2}\cdot\overline{L3}$ | $q_1(t)q_2(t)q_3(t)$ | 0 | $\lambda_1+\lambda_2+\lambda_3$ |

**Table 4.3.** *Unavailability states and their critical transitions*

### 4.7.4. Modeling system unavailability

Of course, the unavailability of the detection system can be determined using equation [4.5]. We could also have determined the unavailability by applying the same approach used in section 4.5, based on the state graph (Figure 4.3) as follows:

$$\overline{\Omega}=L1\cdot\overline{L2}\cdot\overline{L3}+\overline{L1}\cdot\overline{L2}\cdot L3+L1\cdot\overline{L2}\cdot\overline{L3}+\overline{L1}\cdot\overline{L2}\cdot\overline{L3}, \quad [4.25]$$

where $\overline{L\iota}$ describes the states of unavailability of the detector "$i$", $i=1,2,3$ and $\overline{\Omega}$ is the unavailability space.

Because of equation [4.25], we can proceed directly to the evaluation of the unavailability $Q_{\mathrm{sys}}(t)$ of the detection system by:

$$Q_{\mathrm{sys}}(t)=p_1(t)q_2(t)q_3(t)+\ q_1(t)q_2(t)p_3(t) \\ +p_1(t)q_2(t)q_3(t)+\ p_1(t)q_2(t)q_3(t). \quad [4.26]$$

One can always demonstrate that the availability $P_{\mathrm{sys}}(t)$, equation [4.24], and the unavailability $Q_{\mathrm{sys}}(t)$, equation [4.26], always satisfy the normalization condition at each instant $t$, such as:

$$P(t)\ +\ Q(t)\ =1.$$

The evolution over time of the unavailability $Q_{\mathrm{sys}}(t)$ of the detection system is presented in Figure 4.4. We note that the unavailability of the system to carry out its mission increases with time. It starts from ~$10^{-5}$ to reach its asymptotic value of ~$6\times10^{-3}$ in almost 14 h after the detection system is put into service, under the condition of being "as good as new".

It is obvious that an increasing unavailability indicates degradation in the realization of the mission of the detection system. We will discuss this in the conclusion of the chapter.

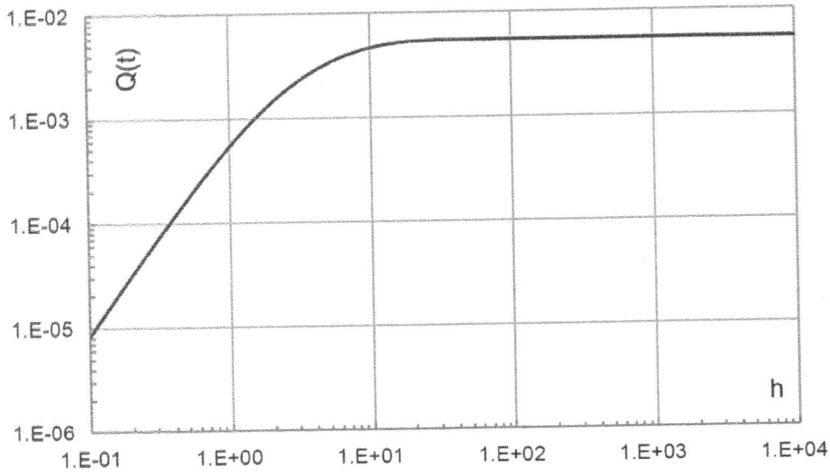


**Figure 4.4.** *Evolution of unavailability of the detection system over time. For a color version of this figure, see www.iste.co.uk/elhami/reliability.zip*

### 4.7.5. Modeling the equivalent failure rate of the system

We will, in this section, focus on the modeling of the equivalent failure rate of the detection system. To this end, we use the state graph in Figure 4.3 to identify the critical transitions that result from constituent detector failure.

The critical failure transitions and the associated states are well identified in Table 4.2. We can therefore model the transition of the detection system from its availability space to unavailability space by:

$$\begin{aligned}\lambda_{sys}(t)P_{sys}(t) = &(\lambda_2+\lambda_3)q_1(t)p_2(t)p_3(t)\\ &+(\lambda_1+\lambda_3)p_1(t)q_2(t)p_3(t)\\ &+(\lambda_1+\lambda_2)p_1(t)p_2(t)q_3(t),\end{aligned} \quad [4.27]$$

where $\lambda_{sys}(t)$ and $P_{sys}(t)$are the equivalent failure rate and the availability of the detection system, respectively.

The evolution of the equivalent failure rate $\lambda_{sys}(t)$ of the detection system is plotted in Figure 4.5. We observe an increase in the failure rate up to its asymptotic value of $10^{-2}/h$.

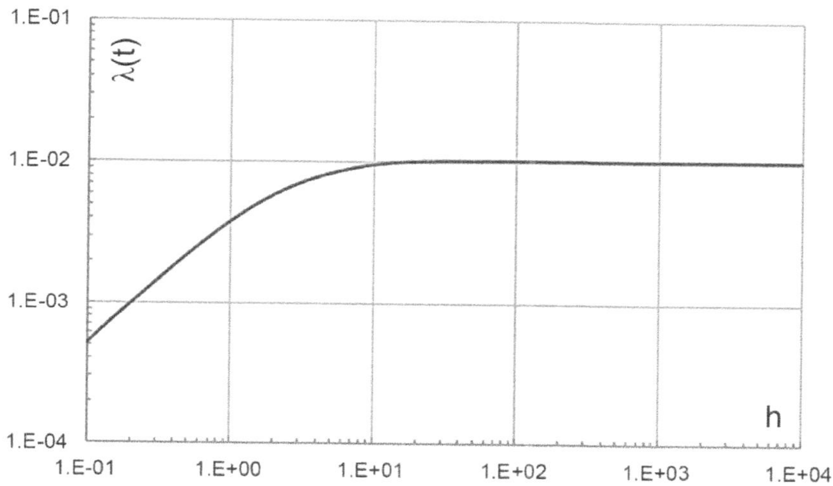


**Figure 4.5.** *The evolution of the equivalent failure rate of the detection system over time. For a color version of this figure, see www.iste.co.uk/elhami/reliability.zip*

### 4.7.6. *The mean time before detection*

The characteristic detection time of the detection system depends on its state at the time of detection. We have identified four probable states for the detection system in its availability space at each instant. The characteristic detection time of the detection system therefore depends on the time.

The characteristic detection time $T(t)$ of the detection system can be determined by:

$$\begin{aligned} T(t)P(t) = & \; p_1(t)p_2(t)p_3(t) \times \tau_2 \\ & + q_1(t)p_2(t)p_3(t) \times \max(\tau_2, \tau_3) \\ & + p_1(t)q_2(t)p_3(t) \times \max(\tau_1, \tau_3) \\ & + p_1(t)p_2(t)q_3(t) \times \max(\tau_1, \tau_2). \end{aligned} \quad [4.28]$$

The evolution of $T(t)$ is presented in Figure 4.6. We observe a growth of $T(t)$ with time up to an asymptotic value around 21 s.

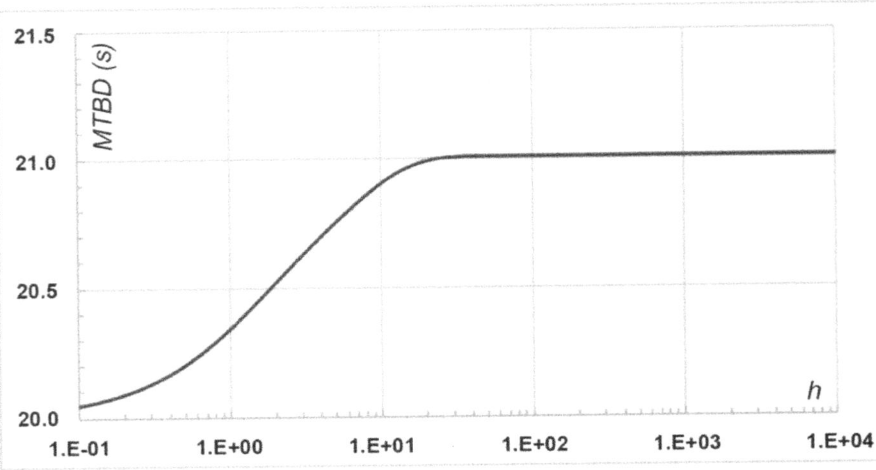


**Figure 4.6.** *Evolution of the mean time before detection (MTBD) of the detection system over time. For a color version of this figure, see www.iste.co.uk/elhami/reliability.zip*

Interestingly, redundancy did not degrade the mean time before detection, despite the increase in detection system downtime and failure rate over time. Nevertheless, Its asymptotic value of 21 s is well below the 30 s of the detector $L3$ and is very close to the 20 s of the detector $L2$.

### 4.7.7. Redundancy effectiveness assessment

Did redundancy really improve the performance of the fire detection function in the workshop?

We will examine the impact of redundancy on the three most relevant indicators: unavailability, failure rate and the mean time to detect fire in the workshop.

Before proceeding in this way, it is necessary to identify the other possible fire detection systems in this case. We can offer three possible systems with two detectors, that is, with no redundancy:

1) $S1$: a system consisting of detectors $L1$, $L2$,

2) $S2$: a system consisting of detectors $L2$, $L3$,

3) $S3$: a system consisting of detectors $L3$, $L1$.

Given the success condition for fire detection, which is that two signals confirm each other, each of the three systems fails if one of the detectors fails. The unavailability is therefore expressed logically by an OR gate. The unavailability of each system is immediately determined by:

$$Q_{ij}(t) = q_i(t) + \big(1 - q_i(t)\big)q_j(t), \quad [4.29]$$

where $i, j(\ i, j = 1,2,3;\ i \neq j)$ are the indices of the two detectors constituting each system.

The unavailability of the redundant system (Figure 4.4) is compared with the unavailability of the three possible non-redundant systems in Figure 4.7.

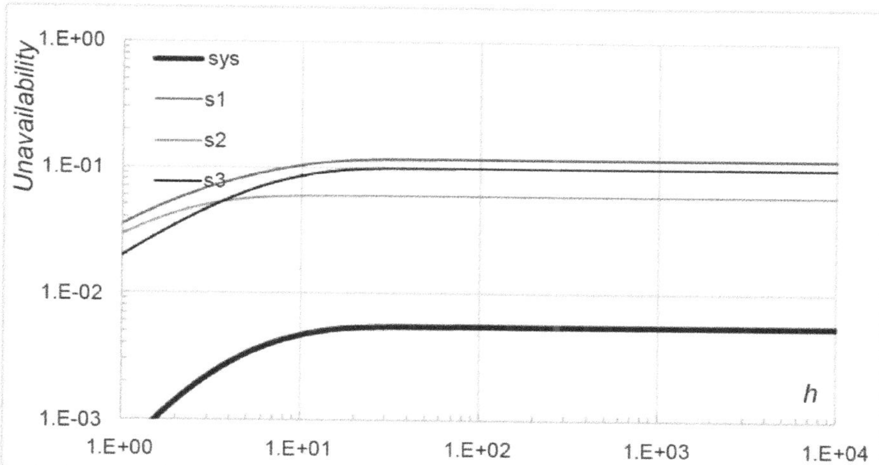


**Figure 4.7.** *Comparison between the unavailability of system 2/3 and the three other possible non-redundant systems ($S_1, S_2, S_3$). For a color version of this figure, see www.iste.co.uk/elhami/reliability.zip*

As for the failure rate $\lambda_{ij}$ of the system consisting of two detectors $i, j$ in gate OR, then:

$$\lambda_{ij} = \lambda_i + \lambda_j. \qquad [4.30]$$

The three systems $(S_1, S_2, S_3)$ are then characterized by three different failure rates, whose respective values are $\lambda_{12} = 1.2^{-2}\text{h}^{-1}, \lambda_{23} = 6.0^{-3}\text{h}^{-1}, \lambda_{31} = 1.0^{-2}\text{h}^{-1}$. As for the failure rate of the redundant system (Figure 4.5), it varies from zero to its asymptotic $1{,}0^{-2}\text{h}^{-1}$ value. Only the system $S_2$ can claim to be better in terms of failure rate, without taking into account either the unavailability or the mean detection time.

The mean detection times $\tau_{ij}$ for each system are determined by:

$$\tau_{ij} = \max\,(\tau_i, \tau_j).$$

The three systems $(S_1, S_2, S_3)$ are then characterized by three different mean detection times, the respective values of which are $\tau_{12} = 20\text{ s}, \tau_{23} = 30\text{ s}, \tau_{31} = 30\text{ s}$. As for the average detection time of the redundant system (Figure 4.6), it varies from 20 s to its asymptotic value 21 s. The system $S_1$ can claim to be slightly better, without considering the downtime in case of failure or the failure rates.

In summary, the redundant system of our case has the lowest unavailability compared to the three other possible fire detection systems. However, its failure rate exceeds that of the system $S_2$with a ratio equal to 67%. Its detection time slightly exceeds that of systems $S_1$with a ratio equal to 5%.

# 5

# System Reliability – Application

System reliability and availability modeling and analysis are among the most common activities in all areas related to systems engineering: system design, maintenance management and optimization, maintenance policy management, periodic tests and operational safety (normal and accidental). In this chapter, we will present a case study using a simple system consisting of five basic components.

The choice of a simple system is intended to avoid inferences from the structural complexity of a system that may obscure the details of the dynamic modeling of the system reliability. It is also intended to encourage the reader to carry out by himself/herself the various steps involved in the process of the dynamic modeling.

This dynamic modeling process goes in the following order: the modeling of the unavailability and the reliability of the elementary components, the construction of the functional space associated with the system, the identification of the states associated with the availability space and those associated with the unavailability space within the functional space, the identification of the critical states, the determination of the equivalent transition rates (failure and repair) and finally the determination of the availability/unavailability and of the reliability/unreliability of the system.

For this didactic case, the elementary components are all assumed to have constant transition rates $(\lambda, \mu)$ over time. This additional simplification is also intended so as not to divert the reader's attention from the main objective of this application: "the mastery of the process of dynamic modeling of the reliability and the availability of systems". However, in the case of a system where the transition rates of the elementary components are time dependent and well defined, the transposition of the modeling process is straightforward.

In this chapter, we will start with a description of the system in section 5.1. In section 5.2, the missions of the system are defined. In section 5.3, the characteristic data of the elementary components $(\lambda, \mu, \gamma)$ are specified. The modeling of the system unavailability and its evaluation are the subjects of sections 5.4 and 5.5, respectively. Thus, the modeling and the assessment of the system availability are the subjects of sections 5.6 and 5.7, respectively. Reliability modeling is discussed in section 5.8, and reliability assessment is presented in section 5.9. At this point, the system is fully determined regarding the system availability and reliability.

Thereafter, we will present two examples of analysis, in section 5.10, to introduce the reader to the activities of reliability analysis of systems that go beyond the simple determination of the availability and the reliability of a system. The two examples given express two different types of questioning: one is of predictive features and the other is of operational ones.

## 5.1. Description of the system

The proposed system and its basic components are schematically presented under the name "water injection system" (SIE) (Figure 5.1).

The SIE is made up of two water supply lines, fed by the same water tank. Each supply line is made up of an electric pump and a motorized valve. The water tank is equipped with pressure and flow control units. The pumps, the valves and the tank are all treated as elementary components[1]. That is to say that they are independent and their characteristic parameters $(\lambda, \mu, \gamma)$ are well defined.

Note that neither the pump unit, the motorized valve nor the tank, with its control units, are elementary components with binary operational mode. However, the simplification of the system is necessary to achieve the didactic objective sought.

1 It should be noted that the pumps, the valves and the tank could have been considered "systems", if their respective constituents had been identified. Our hypothetical knowledge of their characteristic parameters $(\lambda, \mu, \gamma)$ allows us to consider them "elementary components".

A structured set of components is not yet a system. A system is born when missions are assigned to it.

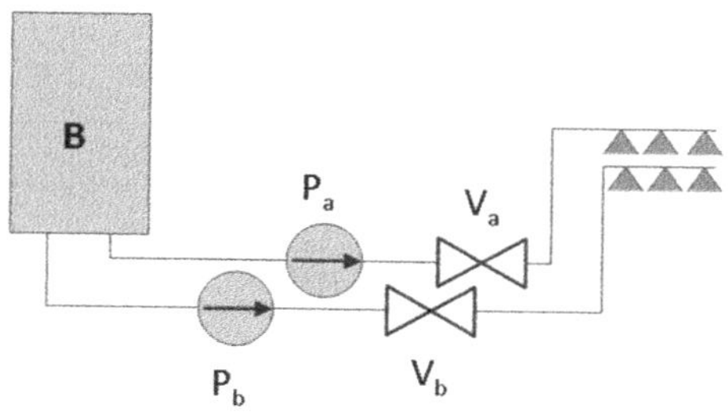


**Figure 5.1.** *Schematic representation of the SIE system*

## 5.2. System missions

The SIE is a safety system. It is designed to inject water to cool a vessel that confines a highly exothermic process, under high temperature and high pressure conditions, in case of an abnormal operating situation.

The system is designed to perform two missions in the event of a deviation from normal operational conditions:

– M1: inject a maximum quantity of water at a well-defined flow rate during the first 18 h of the deviation from the normal operating conditions. The requirement is that both water supply lines must be functioning successfully during the first 18 h.

– M2: between 18 and 60 h, it would suffice that a single supply line keeps in operation.

From now on, the structure of the system is fixed and its missions are well defined. We then need to specify the data characterizing the elementary components.

## 5.3. Basic data

The five basic components are perfectly determined. Their characteristic parameters $(\lambda, \mu, \gamma)$ are well specified (see Table 5.1). The parameters are assumed to be independent of each other and independent of time. Transition rates $(\lambda, \mu)$ are measured in $(h^{-1})$.

Only the pumps show a non-zero refusal to start probability ($\gamma$). However, motorized valves show the highest failure and repair rates, ($\lambda, \mu$).

| # | Component | $\lambda$ ( $h^{-1}$) | $\mu$ ( $h^{-1}$) | $\gamma$ | $q_\infty = \dfrac{\lambda}{\lambda+\mu}$ |
|---|---|---|---|---|---|
| 1 | Electric pump | $1.0^{-4}$ | $1.0^{-1}$ | $8.0^{-5}$ | $10^{-3}$ |
| 2 | Motorized valve | $1.0^{-3}$ | $5.0^{-1}$ | 0 | $2 \times 10^{-3}$ |
| 3 | Tank | $1.0^{-5}$ | $1.0^{-1}$ | 0 | $10^{-4}$ |

**Table 5.1.** *The characteristic parameters of the components*

Regarding the asymptotic unavailability ($q_\infty = \frac{\lambda}{\lambda+\mu}$) of each component, we find that the asymptotic unavailability of the tank is ~$10^{-4}$, about $10^{-3}$ for the pump and about $2 \times 10^{-3}$ for the valve.

## 5.4. System unavailability modeling

The failure of the SIE can be graphically described by a fault tree. Concerning mission $M1$, the corresponding fault tree is a simple "OR" gate on the five components. The fault tree corresponding to mission $M2$ is presented in Figure 5.2.

Note that the analyst could continue the analysis of the failure of each of the components (tank, pumps and valves) down to their most basic failures if these basic failures were identified. The level of decomposition of the failure of a system into more and more elementary failures depends only on the complexity of the system, the availability of elementary failure data and the degree of the expertise of the analyst.

The failure of the SIE can also be formalized by Boolean expressions as follows. If the expressions $\overline{B}, \overline{P_1}, \overline{P_2}, \overline{V_1}, \overline{V_2}$, respectively, describe the unavailability of the tank, the pump 1, the pump 2, the motorized-valve1 and the motorized valve 2, we can express the unavailability of the system for its first mission by $\overline{M1}$, in the interval [0 h, 18 h].

Thus, the set of failure conditions $\overline{M1}$ will be defined by the Boolean expression:

$$\overline{MI} = \bar{B} + (\overline{P_1} + \overline{V_1}) + (\overline{P_2} + \overline{V_2}). \quad [5.1]$$

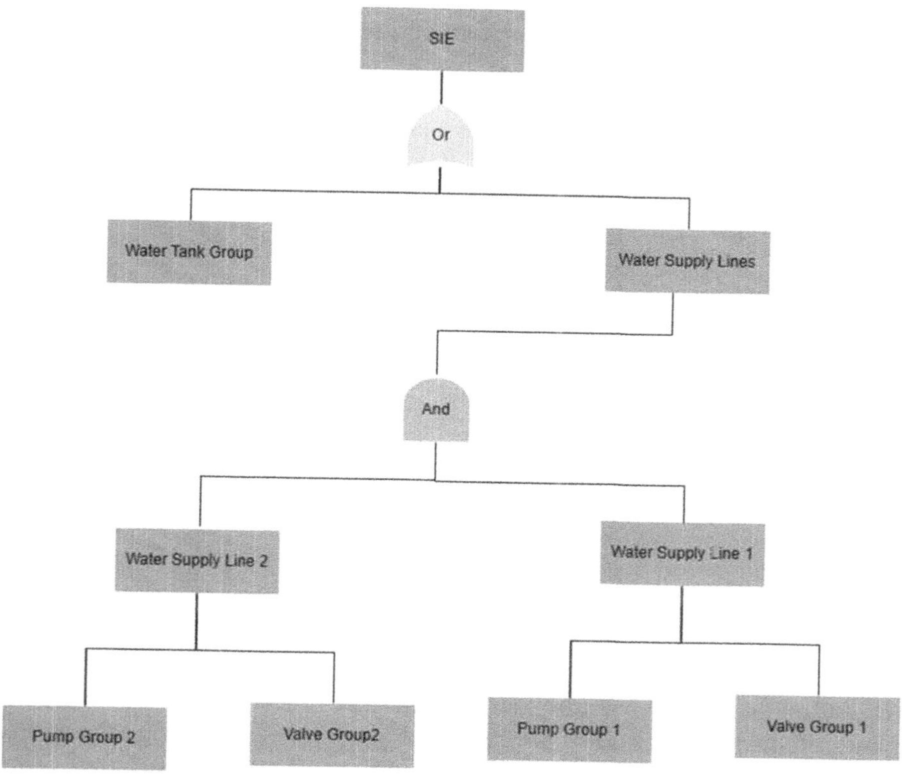


**Figure 5.2.** *SIE failure tree (M2). For a color version of this figure, see www.iste.co.uk/elhami/reliability.zip*

The set of failure conditions $\overline{M2}$ contains all the elementary failures leading to the unavailability of the system regarding its second mission in the interval $[18\text{ h}, 60\text{ h}]$; then, $\overline{M2}$ is defined by the following Boolean expression:

$$\overline{M2} = \bar{B} + (\overline{P_1} + \overline{V_1}) \cdot (\overline{P_2} + \overline{V_2}) = \bar{B} + \overline{P_1} \cdot \overline{P_2} + \overline{P_1} \cdot \overline{V_2} + \overline{V_1} \cdot \overline{P_2} + \overline{V_1} \cdot \overline{V_2} \quad [5.2]$$

where:

$\bar{B}$ designates the unavailability of the water tank and its equipment, $(\overline{P_1} + \overline{V_1})$ designates the unavailability of the first water supply line, $(\overline{P_2} + \overline{V_2})$ designates the unavailability of the second water supply line and the operators "$\cdot$" and " $+$ " are, respectively, the Boolean operators "AND" and "OR".

Equations [5.1] and [5.2] are logical equations in the Boolean sense of the term. Likewise, for the expressions $\bar{B}, \overline{P_1}, \overline{P_2}, \overline{V_1}, \overline{V_2}$, which represent the failures, and the expressions $B, P_1, P_2, V_1, P_2$, which represent the correct functioning of the elementary components.

The expressions of the unavailability in equations [5.1] and [5.2] are described by the minimal cuts of unavailability. The minimal cuts represent the minimal necessary conditions: either to make a system unavailable or to keep the system available (section 1.2).

Because of the Boolean expressions of the unavailability, we can proceed to the determination of the availability and/or the unavailability of the system.

## 5.5. System unavailability assessment

We are interested in the evolution over time of the unavailability of the SIE during its M1 and M2 missions. It should be noted that the SIE starts its mission following the start of the abnormal operation state. During the entire interval of its missions (60 h), only the conditions of its mission change. The system structure undergoes no change during these 60-h intervals: neither structural nor functional. Only the failure/success conditions of the mission assigned to the system change.

During the first mission [0 h, 18 h], the unavailability $\overline{M1}$ described by equation [5.1], one may use one of the equivalent expressions using disjoint cuts, as follows[2]:

$$\overline{M1} = \bar{B} + \mathrm{B} \cdot \overline{P_1} + \mathrm{B} \cdot P_1 \cdot \overline{V_1} + \mathrm{B} \cdot P_1 \cdot V_1 \cdot \overline{P_2} + \mathrm{B} \cdot P_1 \cdot V_1 \cdot P_2 \cdot \overline{V_2}. \qquad [5.3]$$

2 If $S = C1 + C2 + C3 + C4 + \cdots$, then $S^* = C1 + \overline{C1} \cdot C2 + \overline{C1} \cdot \overline{C2} \cdot C3 + \overline{C1} \cdot \overline{C2} \cdot \overline{C3} \cdot C4 + \cdots$ where $S^*$ is an equivalent expression of $S$.

Boolean expressions in [5.1] and [5.3] are equivalent and have the same logical value. But one is expressed in minimal cuts and the other in disjoint (/exclusive) cuts. The Boolean expression in [5.1] can have 120 possible equivalent expressions because it has five minimal cuts (5! = 120). Since these are exclusive cuts, one can directly determine the unavailability $Q1(t)$ of the SIE, for its first mission, by equation [5.4]:

$$Q1(\mathrm{t}) = q_B(t) + p_B(t)q_{P_1}(t) + p_B(t)p_{P_1}(t)q_{V_1}(t) + p_B(t)p_{P_1}(t)p_{V_1}(t)q_{P_2}(t) + p_B(t)p_{P_1}(t)p_{V_1}(t)p_{P_2}(t)q_{V_2}(t), \quad [5.4]$$

where $p_i(t), q_i(t)$are the availability and the unavailability of the elementary component "*i*" (see equation [1.3]).

For the second mission, we also develop an equivalent expression to equation [5.2]:

$$\overline{M2} = \bar{B} + B \cdot (\overline{P_1} + P_1 \cdot \overline{V_1}) \cdot (\overline{P_2} + P_2 \cdot \overline{V_2}). \quad [5.5]$$

The unavailability $Q2(t)$ corresponding to M2 will be determined by:

$$Q2(t) = q_B(t) + p_B(t) \cdot \left[(q_{P_1}(t) + p_{P_1}(t)q_{V_1}(t)) \cdot (q_{P_2}(t) + p_{P_2}(t)q_{V_2}(t))\right]. \quad [5.6]$$

The evolutions over time of the unavailability, $Q1(t)$ and $Q2(t)$, are plotted in Figure 5.3 by the solid red and blue lines, respectively.

We make these two observations:

– the system reaches its asymptotic plateau after ~20 h of its solicitation;

– the asymptotic unavailability of the system during the second mission is very close to the asymptotic unavailability of the tank. The conclusion is that the unavailability of the system is controlled by the unavailability of the tank during its second mission.

The determination of unavailability $Q1(t)$ indirectly determines the availability $P1(t)$ of the system because the two quantities are complementary.

However, availability can also be determined from Boolean expressions of availability. For educational purposes, we will determine the availability of the system from the Boolean expressions of the unavailability starting from equations [5.1] and [5.2].

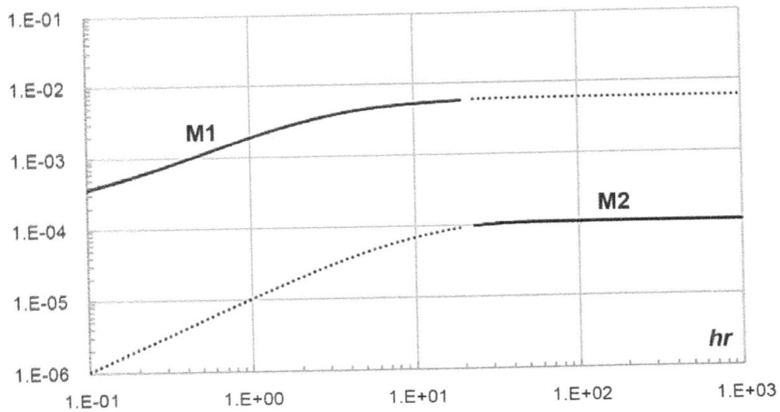


**Figure 5.3.** *The evolution over time of the unavailability of the SIE during its two missions M1 [0 h, 18 h] and M2 [18 h, 60 h]. For a color version of this figure, see www.iste.co.uk/elhami/reliability.zip*

## 5.6. System availability modeling

Obviously, we could have described the availability instead of the failure of the system, using the reliability block diagram[3] instead of the failure tree. We can also deduce the Boolean expression of the availability directly from the Boolean expressions of the unavailability (equations [5.1] and [5.2]).

Then, we can express the availability during the first mission, $M1$, by the Boolean expression:

$$M1 = \overline{(\overline{M1})} = \overline{\left(\bar{B} + (\overline{P_1} + \overline{V_1}) + (\overline{P_2} + \overline{V_2})\right)}.$$

The Boolean expression of availability is therefore:

$$M1 = B \cdot (P_1 \cdot V_1) \cdot (P_2 \cdot V_2). \tag{5.7}$$

3 The block diagram is to be defined.

The expression of the availability of the system during the first phase therefore contains only one minimal cut of 5th order.

As for the expression of the availability $M2$ during the second mission, it is defined by:

$$M2 = \overline{(\overline{M2})} = \overline{(\bar{B} + [(\bar{P_1} + \bar{V_1}) \cdot (\bar{P_2} + \bar{V_2})])}.$$

Thus, the Boolean expression of the availability $M2$ is therefore:

$$M2 = B \cdot \left((P_1 \cdot V_1) + (P_2 \cdot V_2)\right) = B \cdot P_1 \cdot V_1 + B \cdot P_2 \cdot V_2. \quad [5.8]$$

The Boolean expression of availability $M2$ in equation [5.8] contains two minimal cuts of order three each. One of the expressions equivalent to the Boolean expression [5.8] is as follows:

$$M2 = B \cdot P_1 \cdot V_1 + B \cdot \overline{(P_1 \cdot V_1)} \cdot P_2 \cdot V_2. \quad [5.9]$$

The Boolean expression of the availability in equation [5.9] makes it possible to directly determine the availability of the system during its second phase because it expresses the availability in disjoint cuts.

We will leave to the readers the verification of the complementarity between equations [5.1 ] and [5.2] relating to the two missions of the system and equations [5.7] and [5.8], respectively. The verification of the complementarity can be carried on logically and/or numerically.

## 5.7. System availability assessment

The Boolean expression of equation [5.7] is usable in its state without recourse to equivalent expressions with disjoint cuts because it contains only one cut. Then, the availability of the system $P_1(t)$, during the first phase, is directly determined by:

$$P1(t) = p_B(t) \cdot p_{P_1}(t) \cdot p_{V_1}(t) \cdot p_{P_2}(t) \cdot p_{V_2}(t). \quad [5.10]$$

As for the Boolean expression of equation [5.9], it makes it possible to directly determine the availability of the system during the second mission, as follows:

$$P2(t) = p_B(t) \cdot \Big[p_{P_1}(t) \cdot p_{V_1}(t) + \Big(1 - p_{P_1}(t) \cdot p_{V_1}(t)\Big) \cdot p_{P_2}(t) \cdot p_{V_2}(t)\Big]. \quad [5.11]$$

System analysts do often seek to determine and analyze the unavailability more than the availability of systems. The analysis of unavailability is often more significant and discriminant than the availability analysis.

Note that the Boolean expression of the failure $\overline{M1}$, equation [5.1]), contains five distinct minimal cuts. The cuts are all of order one. However, the Boolean expression $\overline{M2}$, equation [5.2]), also contains five minimal cuts, including one of first order and four of second order.

If we examine the Boolean expressions of the availability of the system, we find that the Boolean expression of the availability [5.7] contains only one minimal cut of order five. However, the Boolean expression [5.8] contains two minimal cuts of order three each.

We will leave to the readers the verification of the complementarity between equations [5.4] and [5.6] relating to the two missions of the system and equations [5.10] and [5.11].

## 5.8. Reliability modeling

For a safety system such as the SIE system, the modeling and the evaluation of its reliability for all its missions are of capital interest. For a purely didactic objective, we are going to apply two very distinct approaches, but the approaches have the same logical basis. One will use the logical expression of the SIE's availability for the first mission. The other will use the system state graph for the second mission.

To assess the reliability of the SIE for the first mission, we will use equation 5.7:

$$M1 = B \cdot P_1 \cdot V_1 \cdot P_2 \cdot V_2.$$

Being a logical expression with only one minimal cut, we can use the same condition for reliability. This can be formulated by "the system is reliable if and only if all the elementary components are also reliable, during the time interval of interest". The reliability $F1(\mathrm{t})$ of the SIE during its first M1 mission (18 h) can then be determined by:

$$F1(t) = f_B(t) f_{P_1}(t) f_{V_1}(t) f_{P_2}(t) f_{V_2}(t). \qquad [5.12a]$$

This will give:

$$F1(\mathrm{t}) = e^{-(\lambda_B+\lambda_{P_1}+\lambda_{V_1}+\lambda_{P_2}+\lambda_{V_2})t}. \qquad [5.12b]$$

Unsurprisingly, equation [5.13] expresses the fact that the system behaves like a serial system during M1 mission. The failure of a single elementary component makes the whole system fail. This corresponds obviously to the five minimal cuts, which are all of order one, in failure case. The equivalent failure rate of the system $\Lambda_{sys}1$ during the first mission (0 h,18 h) can then be determined:

$$\Lambda_{sys}1 = \lambda_B + \lambda_{P_1} + \lambda_{V_1} + \lambda_{P_2} + \lambda_{V_2}. \qquad [5.13]$$

Therefore, $\Lambda_{sys}1$ is constant over time and is equal to $2.2^{-3}/\mathrm{h}$. The reliability of the system $F1(t = 18)$ over the interval of the first mission (18 h) is evaluated at $9.61^{-1}$.

As for the reliability of the system during the second mission [18 h–60 h], we are going to determine the equivalent failure rate of the system by using its state graph. Specifically, we need to identify the critical transitions using the state graph. We will use equation [5.8] as a logical condition to identify the availability states of the system during its second mission [18 h–60 h]. Next, we will identify the critical transitions (section 2.2) to determine the equivalent failure rate of system $\Lambda_{sys}2$ as a function of time. The examination of the availability space of the system has enabled us to identify a total of seven states that satisfy the logical conditions in equation [5.8], out of the 32 ($2^5$) states defining the functional space $\Omega$ of the system. These seven states, which constitute the availability space and their critical transitions, are presented on the system state graph in Figure 5.4. The failure states, 25 in number, are not detailed in the state graph for the sake of easy readability of the graph.

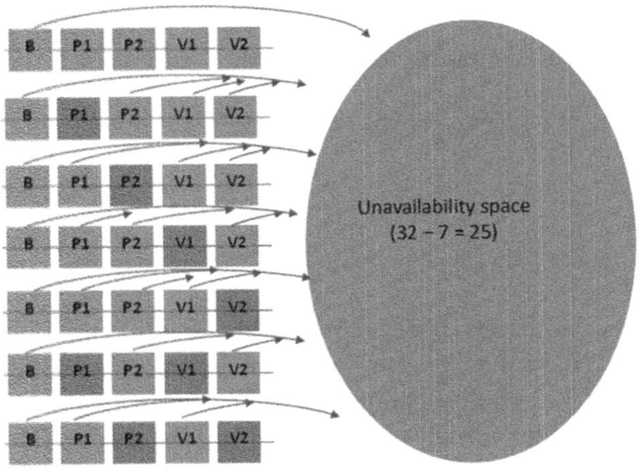


**Figure 5.4.** *The SIE state graph with its seven-state availability space (green square = operation, red square = unavailable). For a color version of this figure, see www.iste.co.uk/elhami/reliability.zip*

We then evaluate the equivalent failure rate of the system because of the data related to the critical transitions (in Table 5.2) and to the application of the following equation.

$$\begin{aligned}\Lambda_{sys}2(t)\cdot P2(t) = &\ (1.0^{-5})[p_B(t)p_{P1}(t)p_{P2}(t)p_{V1}(t)p_{V2}(t)]\\ &+ (1.1^{-3})[p_B(t)q_{P1}(t)p_{P2}(t)p_{V1}(t)p_{V2}(t)]\\ &+ (1.1^{-3})[p_B(t)p_{P1}(t)q_{P2}(t)p_{V1}(t)p_{V2}(t)]\\ &+ (1.1^{-3})[p_B(t)p_{P1}(t)p_{P2}(t)q_{V1}(t)p_{V2}(t)]\\ &+ (1.1^{-3})[p_B(t)p_{P1}(t)p_{P2}(t)p_{V1}(t)q_{V2}(t)]\\ &+ (1.1^{-3})[p_B(t)q_{P1}(t)p_{P2}(t)q_{V1}(t)p_{V2}(t)]\\ &+ (1.1^{-3})[p_B(t)p_{P1}(t)q_{P2}(t)p_{V1}(t)q_{V2}(t)],\end{aligned} \quad [5.14]$$

where $P2(t)$ is already determined by equation [5.11].

The equivalent failure rate during the second mission [18 h–60 h] is time dependent (equation [5.14]). The evolution over time of the equivalent failure rate during the M2 mission is plotted in Figure 5.5.

The evolution over time of the equivalent failure rate during the second mission is also presented numerically in Table 5.3. We find that the SIE reaches its asymptotic value around 30 h and that the asymptotic value of the equivalent failure rate is equal to $\sim 1.66^{-5}/\text{h}$ (Figure 5.5). We can therefore determine the time integral of the equivalent failure rate numerically, as shown in Table 5.3. The reliability of the system during its second mission is therefore equal to ~0.9993.

| $E_i$ | $B$ | $P1$ | $P2$ | $V1$ | $V2$ | CT | Critical transition rate | $\Lambda_i$ (/h) | prob. $\mathrm{Pr}_i(t)$ |
|---|---|---|---|---|---|---|---|---|---|
| 1 | 0 | 0 | 0 | 0 | 0 | 1 | $\Lambda 1 = \lambda_B$ | $1.0^{-5}$ | $p_B p_{P1} p_{P2} p_{V1} p_{V2}$ * |
| 2 | 0 | 1 | 0 | 0 | 0 | 3 | $\Lambda 2 = \lambda_B + \lambda_{P_2} + \lambda_{V_2}$ | $1.1^{-3}$ | $p_B q_{P1} p_{P2} p_{V1} p_{V2}$ |
| 3 | 0 | 0 | 1 | 0 | 0 | 3 | $\Lambda 3 = \lambda_B + \lambda_{P_1} + \lambda_{V_1}$ | $1.1^{-3}$ | $p_B p_{P1} q_{P2} p_{V1} p_{V2}$ |
| 4 | 0 | 0 | 0 | 1 | 0 | 3 | $\Lambda 4 = \lambda_B + \lambda_{P_2} + \lambda_{V_2}$ | $1.1^{-3}$ | $p_B p_{P1} p_{P2} q_{V1} p_{V2}$ |
| 5 | 0 | 0 | 0 | 0 | 1 | 3 | $\Lambda 5 = \lambda_B + \lambda_{P_1} + \lambda_{V_1}$ | $1.1^{-3}$ | $p_B p_{P1} p_{P2} p_{V1} q_{V2}$ |
| 6 | 0 | 1 | 0 | 1 | 0 | 3 | $\Lambda 6 = \lambda_B + \lambda_{P_2} + \lambda_{V_2}$ | $1.1^{-3}$ | $p_B q_{P1} p_{P2} q_{V1} p_{V2}$ |
| 7 | 0 | 0 | 1 | 0 | 1 | 3 | $\Lambda 7 = \lambda_B + \lambda_{P_1} + \lambda_{V_1}$ | $1.1^{-3}$ | $p_B p_{P1} q_{P2} p_{V1} q_{V2}$ |

* $p_i, q_i$ are, respectively, the availability and unavailability of the elementary components "$i$" (equation [1.3]).

**Table 5.2.** *Availability states and associated critical transitions*

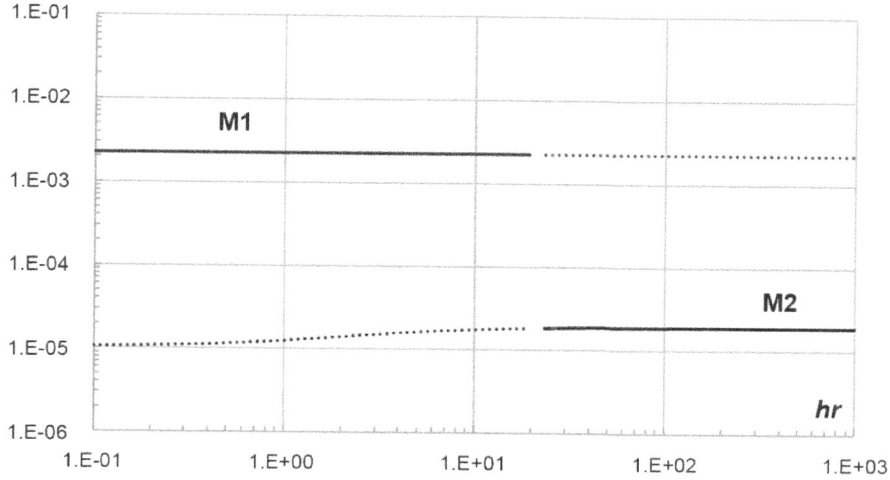


**Figure 5.5.** *The equivalent SIE failure rate over time during the first and second missions. For a color version of this figure, see www.iste.co.uk/elhami/reliability.zip*

| $I$ | Time (h) | $\Lambda 2(t)(\mathrm{h}^{-1})$ | $\int_{t_{i-1}}^{t_i} \Lambda(t)dt$ | Comments |
|---|---|---|---|---|
| 1 | 1.80E+01 | 1.62E–05 | | |
| 2 | 1.96E+01 | 1.63E–05 | 2.62E–05 | |
| 3 | 2.14E+01 | 1.63E–05 | 2.86E–05 | |
| 4 | 2.33E+01 | 1.64E–05 | 3.13E–05 | |
| 5 | 2.54E+01 | 1.64E–05 | 3.41E–05 | |
| 6 | 2.76E+01 | 1.64E–05 | 3.73E–05 | |
| 7 | 3.01E+01 | 1.65E–05 | 4.07E–05 | |
| 8 | 3.28E+01 | 1.65E–05 | 4.44E–05 | |
| 9 | 3.57E+01 | 1.65E–05 | 4.84E–05 | |
| 10 | 3.89E+01 | 1.65E–05 | 5.28E–05 | |
| 11 | 4.24E+01 | 1.65E–05 | 5.76E–05 | |
| 12 | 4.62E+01 | 1.65E–05 | 6.28E–05 | |
| 13 | 5.03E+01 | 1.66E–05 | 6.84E–05 | |
| 14 | 5.48E+01 | 1.66E–05 | 7.46E–05 | |
| 15 | 5.98E+01 | 1.66E–05 | 8.13E–05 | |
| $\Phi = \int_{18}^{60} \Lambda(t)dt = \sum_{i=1}^{15} \int_{t_{i-1}}^{t_i} \Lambda(t)dt$ | | | 6.88E–04 | |
| Reliability $(M2) = e^{-\Phi}$ | | | 0.9993 | |

**Table 5.3.** *Equivalent failure rate over time, for the second mission*

## 5.9. System reliability assessment

The total reliability of the SIE over the interval [0,18], which covers the first mission, is equal to $9.61^{-1}$, and the reliability of the system over the interval [18, 60] of the second mission is equal to $9.993^{-1}$.

The reliability of the system $R_{sys}(60)$ over the interval [0, 60], which covers these two missions, is therefore:

$$R_{sys}(60) = 9.61^{-1} \times 9.993^{-1} = 9.6^{-1}.$$

## 5.10. Analysis of the relative importance of components

The unavailability of a system results from the unavailability of its components under certain conditions well be defined by the minimal conditions of unavailability (the minimal cuts). The contribution of a given component to system unavailability varies from component to another and from an instant to another. System analysts then seek to determine the weight of each component in the unavailability of the system at each instant. This "weight" can be measured by the "unavailability importance index".

We will review in this section two classes of importance indices in unavailability. One of the classes is of an a priori (predictive) nature and the other a posteriori (operational) nature.

### 5.10.1. *The predictive importance index*

The predictive importance index makes it possible to identify the component and/or the group of components whose unavailability will have the greatest probability of making the system unavailable. The predictive index of importance $I_p$ is measured by "the probability that the system becomes unavailable, knowing that one of the components or a group of components is unavailable, at the instant $t$". This analysis is made in anticipation of possible unavailability of the system. It helps guiding preventive maintenance actions regarding the components or the groups of elementary components.

Let us take the previous case of the SIE system during its second mission. The Boolean expression for system unavailability is defined in equation [5.2], and system unavailability $Q2(t)$ during its second mission is determined by equation [5.6]. We wish to determine the predictive importance index of the unavailability of a supply line $I_{pL1}$.

The unavailability of the supply line is defined by the $\overline{L1}$ following Boolean expression:

$$\overline{L1} = \overline{P_1} + \overline{V_1} = \overline{P_1} + P_1 \cdot \overline{V_1}. \quad [5.15]$$

The unavailability $Q_{\overline{L1}}(t)$ of the line is determined by the following equation:

$$Q_{\overline{L1}}(t) = q_{P_1} + p_{P_1} q_{v_1}. \qquad [5.16]$$

The event of having the line $\overline{L1}$ and the system both unavailable at instant $t$ during the second mission is described by the following Boolean expression:

$$\begin{aligned}\overline{L1} \cdot \overline{M2} &= (\overline{P_1} + \overline{V_1}) \cdot \left(\bar{B} + (\overline{P_1} + \overline{V_1}) \cdot (\overline{P_2} + \overline{V_2})\right) \\ &= (\overline{P_1} + \overline{V_1}) \cdot (\bar{B} + \overline{P_2} + \overline{V_2}).\end{aligned}$$

Then, we can put the previous expression under one of its equivalent expressions as follows:

$$\overline{L1} \cdot \overline{M2} = (\overline{P_1} + P_1 \cdot \overline{V_1}) \cdot (\bar{B} + B \cdot \overline{P_2} + B \cdot P_2 \cdot \overline{V_2}). \qquad [5.17]$$

The probability $Q_{\overline{L1}\cdot\bar{S}}(t)$ of having a line and the system unavailable $(\overline{L1} \cdot \overline{M2})$ at instant $t$ is determined by:

$$\begin{aligned}Q_{\overline{L1}\cdot\bar{S}}(t) &= \left(q_{P_1}(t) + p_{P_1}(t) q_{V_1}(t)\right) \\ &\cdot \left(q_B(t) + p_B(t) q_{P_2}(t) + p_B(t) p_{P_2}(t) q_{V_2}(t)\right).\end{aligned} \qquad [5.18]$$

The predictive importance index $I_{pL1}(t)$ of a line $L1$ is measured by "the probability that the system becomes unavailable given that a water supply line is unavailable". Thus $I_{pL1}(t)$ is determined by:

$$I_{pL1}(t) = \frac{Q_{\overline{L1}\cdot\bar{S}}(t)}{Q_{\overline{L1}}(t)} = q_B(t) + p_B(t) q_{P_2}(t) + p_B(t) p_{P_2}(t) q_{V_2}(t), \qquad [5.19]$$

where $Q_{\overline{L1}\cdot\bar{S}}(t)$, $Q_{\overline{L1}}(t)$ and $I_{pL1}(t)$ are, respectively, determined by equations [5.17], [5.18] and [5.19].

The evolution over time of the index $I_{pL1}(t)$ for the SIE is plotted in Figure 5.6. It is found that the predicted importance of a water supply line in the eventual unavailability of the system increases slightly with time. The importance index reaches its asymptotic value 40 h after the start of the abnormal operation phase, that is, 20 h before the end of the second mission.

Equation [5.19] reveals that if one of the water supply lines is unavailable, the unavailability of the system is driven by the unavailability of the tank or the unavailability of the second supply line. This conclusion is obvious because of the simplicity of the system. It would not have been as obvious if the system was more complex.

Thus, we can arrive at the same conclusion without the aid of Boolean algebra, due to the simplicity of the system. However, equation [5.19] makes it possible to classify systematically the components according to their importance regarding the management of future preventive maintenance actions within the optic of minimizing the future system unavailability profile with time. That would equally include other predictive activities such as the periodic tests and control, system health monitoring, replacing/ upgrading components, and certification.

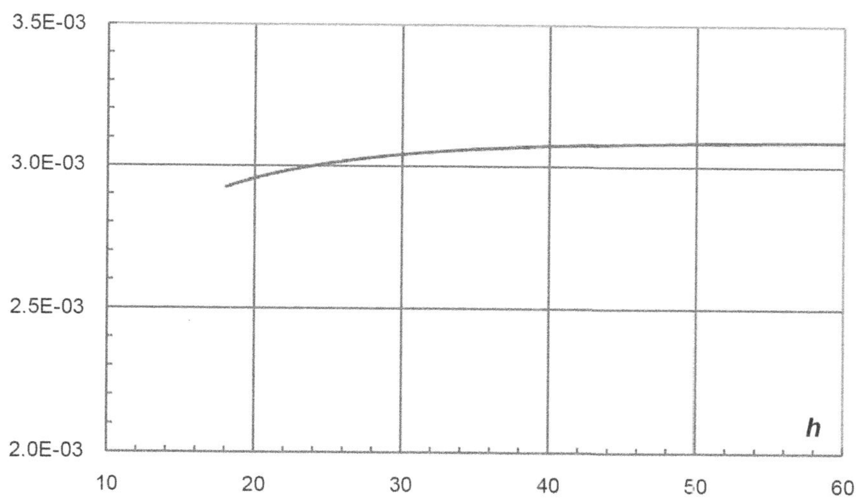


**Figure 5.6.** *The evolution over time of the predictive importance index in unavailability of a supply line, vis-à-vis the unavailability of the SIE system. For a color version of this figure, see www.iste.co.uk/elhami/reliability.zip*

## 5.10.2. *The operational importance index*

The operational importance index in unavailability is, on the other hand, the answer to the concern to identify the component and/or the group of components that are most likely unavailable, when a system is unavailable at the instant $t$. The operational unavailability index $I_o$ is measured by "the probability that a component or a group of components is unavailable, knowing that the system is unavailable at the instant $t$". The issue here is to classify the components to be maintained given that the system is already unavailable.

Let us continue with the SIE system during its second mission. We wish to determine the operational importance index in unavailability of the motorized valves group $\overline{VM}$, vis-à-vis the unavailability of the system. The unavailability of the motorized valves group can be expressed in three different and arbitrary ways:

1) the unavailability of the two system valves;

2) the unavailability of one, but not the other;

3) the unavailability of at least one valve.

We opt for the third configuration, given its didactic interest that we will discuss further. The unavailability of the motorized valves group $\overline{VM}$ will therefore be defined, arbitrarily (at least one valve is unavailable), by the following Boolean expression:

$$\overline{VM} = \overline{V_1} + \overline{V_2} = \overline{V_1} + V_1 \cdot \overline{V_2}. \qquad [5.20]$$

The probability $Q_{\overline{VM}}(t)$ of having at least one of the valves unavailable as described by equation [5.20] will be determined by:

$$Q_{\overline{VM}}(t) = q_{V_1}(t) + p_{V_1}(t) q_{V_2}(t). \qquad [5.21]$$

We will call the probability $Q_{\overline{VM}}(t)$ "motorized valves group unavailability", acknowledging our misuse of the term "unavailability", from a "systems engineering" point of view.

The set of events resulting from the intersection of "the unavailability of the motorized valve group and the unavailability of the system, in the second mission" will therefore be defined by:

$$\overline{VM} \cdot \overline{M2} = (\overline{V_1} + \overline{V_2}) \cdot \left(\bar{B} + (\overline{P_1} + \overline{V_1}) \cdot (\overline{P_2} + \overline{V_2})\right) = \bar{B} \cdot (\overline{V_1} + \overline{V_2}) + \overline{V_1} \cdot (\overline{P_2} + \overline{V_2}) + \overline{V_2} \cdot (\overline{P_1} + \overline{V_1}).$$

Then, we can put the previous expression under one of its equivalent expressions, as follows:

$$\overline{VM} \cdot \overline{M2} = (\bar{B} \cdot \overline{V_1 \cdot V_2}) + B \cdot \bar{V_1} \cdot (\overline{P_2.V_2}) + B \cdot V_1 \cdot \overline{V_2} \cdot \overline{P_1}. \quad [5.22]$$

The probability $Q_{\overline{VM}\cdot\bar{S}}(t)$ of having the motorized valves group and the system unavailable $(\overline{VM} \cdot \overline{M2})$, as described in equation [5.22], is determined by:

$$Q_{\overline{VM}\cdot\bar{S}}(t) = q_B(t) \cdot \left(1 - p_{V_1}(t)p_{V_2}(t)\right) + p_B(t) \cdot q_{V_1}(t) \cdot (1 - p_{P_2}(t)p_{V_2}(t)) + p_B(t)p_{V_1}(t)q_{V_2}(t)q_{P_1}(t). \quad [5.23]$$

The operational importance index $I_{oGM}(t)$ in unavailability of the motorized valves group, measured by “the probability that the motorized valves group is unavailable, given that the system is unavailable at the instant $t$”, is determined by:

$$I_{oGM}(t) = \frac{Q_{\overline{VM}\cdot\bar{S}}(t)}{Q2(t)}, \quad [5.24]$$

where $Q2(t)$ and $Q_{\overline{VM}\cdot\bar{S}}(t)$ are, respectively, determined by equations [5.6] and [5.23].

We will proceed in the same way to determine the operational importance in the unavailability of the electric pumps group. Following the same reasoning as for the motorized valves group, the unavailability of the electric pumps group $\overline{PE}$ will be defined by the following Boolean expression:

$$\overline{PE} = \overline{P_1} + \overline{P_2} = \overline{P_1} + P_1 \cdot \overline{P_2}. \quad [5.25]$$

The probability $Q_{\overline{PE}}(t)$ of occurrence of the events described by equation [5.25] will be determined by:

$$Q_{\overline{PE}}(t) = q_{P_1}(t) + p_{P_1}(t)q_{P_2}(t). \quad [5.26]$$

By convention, we will call the probability $Q_{\overline{PE}}(t)$ “the unavailability of the electric pump group”.

The set of events resulting from the intersection of "the unavailability of the electric pump group and the unavailability of the system in the second mission" will therefore be defined by the following expression:

$$\overline{PE}\cdot\overline{M2} = (\overline{P_1}+\overline{P_2})\cdot\left(\bar{B}+(\overline{P_1}+\bar{V_1})\cdot(\overline{P_2}+\overline{V_2})\right) = \bar{B}\cdot(\bar{V_1}+\overline{V_2})+\bar{V_1}\cdot(\overline{P_2}+\overline{V_2})+\overline{V_2}\cdot(\overline{P_1}+\bar{V_1}).$$

Then, we can put the previous expression under one of its equivalent expressions as follows:

$$\overline{PE}\cdot\overline{M2} = (\bar{B}\cdot\overline{P_1\cdot P_2}) + B\cdot\overline{P_1}\cdot(\overline{P_2.V_2}) + B\cdot P_1\cdot\overline{P_2}\cdot\bar{V_1}. \quad [5.27]$$

The probability $Q_{\overline{PE}\cdot\bar{S}}(t)$ of having the electric pump group and the system unavailable $(\overline{PE}\cdot\overline{M2})$ is determined by:

$$Q_{\overline{PE}\cdot\bar{S}}(t) = q_B(t)\cdot\left(1-p_{p_1}(t)p_{p_2}(t)\right) + p_B\cdot q_{p_1}\cdot(1-p_{P_2}p_{V_2}) + p_B(t)p_{p_1}(t)q_{p_2}(t)q_{v_1}(t). \quad [5.28]$$

The operational importance index in the unavailability of the electric pump group is thus determined by "the probability that the electric pump group is unavailable, knowing that the system is unavailable at the moment $t$", $I_{oPE}(t)$ is therefore determined by:

$$I_{oPE}(t) = \frac{Q_{\overline{PE}\cdot\bar{S}}(t)}{Q2(t)}, \quad [5.29]$$

where $Q2(t)$ and $Q_{\overline{PE}\cdot\bar{S}}(t)$ are, respectively, determined by equations [5.6] and [5.28].

Finally, we will proceed in the same way to determine the operational importance index in the unavailability of the tank group. The set of events resulting from the intersection of "the unavailability of the tank group and the unavailability of the system in the second mission" will therefore be defined by the following expression:

$$\bar{B}\cdot\overline{M2} = \bar{B}\cdot\left(\bar{B}+(\overline{P_1}+\bar{V_1})\cdot(\overline{P_2}+\overline{V_2})\right) = \bar{B}.$$

The probability $Q_{\bar{B}\cdot\bar{S}}(t)$ of having the tank group and the system unavailable $(\bar{B}\cdot\overline{M2})$ is determined by:

$$Q_{\overline{PE}\cdot\bar{S}}(t) = q_B(t). \quad [5.30]$$

The operational importance index in unavailability of the tank group, $I_{oB}(t)$, is therefore determined by:

$$I_{oB}(t) = \frac{q_B(t)}{Q2(t)}, \qquad [5.31]$$

where $Q2(t)$ is determined by equation [5.6].

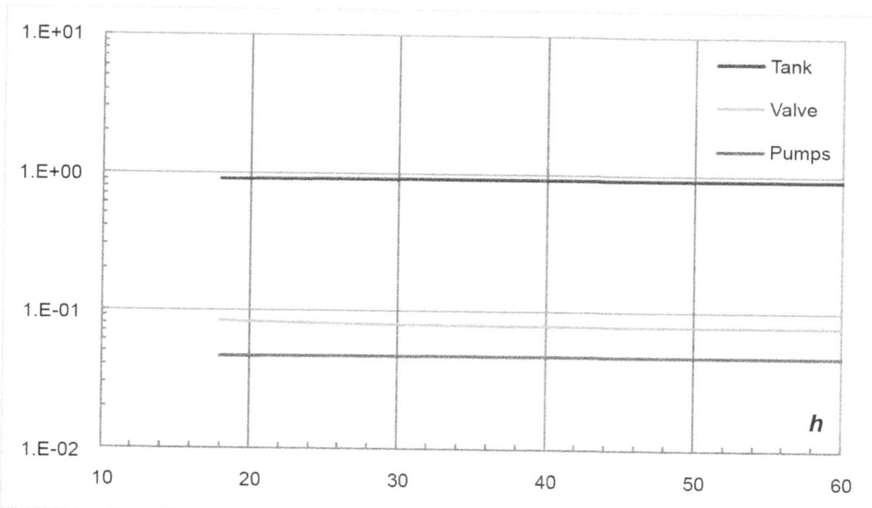


**Figure 5.7.** *Evolution over time of the operational indices of importance in unavailability for the pump, valve and tank groups, given the unavailability of the system. For a color version of this figure, see www.iste.co.uk/elhami/reliability.zip*

Figure 5.7 shows that in case of unavailability of the system, the possible causes of the unavailability are in order of importance: the tank (in most cases), then the valve group and then the pump group (well distanced). This diagnosis allows effective operational management of the interventions of the maintenance teams and their logistics resources in case of a system failure.

## 5.11. Conclusion

Dynamic modeling of a multistate system requires detailed knowledge and description of the system operational modes and its operational states.

Often, in the case of multistate systems with a very large number of states, our computational capacity becomes quickly saturated and the full knowledge of the functional details of each state loses its analytical interest. Very often also, the majority of large systems can be broken down into several independent medium-sized systems. The possibility of decomposing a large complex system into independent medium-size systems may be exploited as a systemic binarization process can make possible the dynamic modeling of large complex systems and their characterization $(\lambda, \mu, \gamma)$.

# 6

# Sequential Events – Modeling and Analysis

## 6.1. Introduction

The analysis of sequential events is a major topic in the analysis of accidents and systems' operational safety. Systems analysts must identify the sequences of success and failures (systemic, human, procedural, organizational, etc.) due to initiating hazardous events at the earlier phases of system design. They seek to eliminate/reduce the occurrence of the feared sequences to interrupt the order of the occurrence of the events in the sequences, and/or to mitigate their consequences.

Analysts thus use several analysis tools, including the event tree analysis. Failure tree analysis considers only failures and does not take into account the sequential aspect of the occurrence of the events. Unlike the failure tree analysis, the event tree analysis considers both the success and the failure as well as the order of occurrence of events.

A set of events $\{a, b, c\}$ can generate six possible sequences (3!=6): $\{a \to b \to c\}$, $\{a \to c \to b\}$, $\{b \to a \to c\}$, $\{b \to c \to a\}$, $\{c \to a \to b\}$ and $\{c \to b \to a\}$. Each of these sequences has eight possible scenarios ($2^3$), but with different occurrence probabilities.

One or more of these scenarios can represent some hazards by their consequences. For the feared scenarios, we need to determine their likelihood, frequency, and consequences.

Analysts often use event tree graphs for the visualization of the sequential occurrence of events. In Figure 6.1, we present an event tree that describes all the possible scenarios of one of the sequences relative to the three independent events $\{a, b, c\}$.

In Figure 6.1, an upward arrow indicates the occurrence of an event, and a downward arrow indicates the non-occurrence. With a sequence $\{a \rightarrow b \rightarrow c\}$ comprising three events, we effectively expect eight possible scenarios defining the probability space $\Omega^3$ of the sequence. As we are interested in the sequence described by $\{a \rightarrow b \rightarrow c\}$, this implies that the events $\{a, b, c\}$ occur respectively in the intervals $[0, t_1]$, $[t_1, t_2]$ and $[t_2, t_3]$ within a total observation interval equal to $T$, such that $(0 \leq t_1 \leq t_2 \leq t_3 \leq T)$.

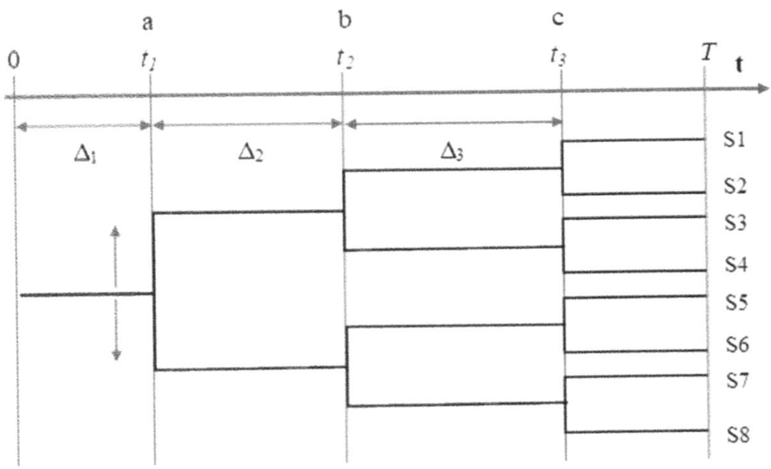


**Figure 6.1.** *Event tree with three independent events. For a color version of this figure, see www.iste.co.uk/elhami/reliability.zip*

The possible scenarios are $S1(a, b, c)$, $S2(a, b, \bar{c})$, $S3(a, \bar{b}, c)$, $S4(a, \bar{b}, \bar{c})$, $S5(\bar{a}, b, c)$, $S6(\bar{a}, b, \bar{c})$, $S7(\bar{a}, \bar{b}, c)$, and $S8(\bar{a}, \bar{b}, \bar{c})$, where the occurrence of an event is marked with "$a$" and the non-occurrence is marked with "$\bar{a}$". Each of the eight scenarios is associated with a probability of occurrence $P(S_i)$, $(i = 1,2,3,4,5,6,7,8)$. The possible scenarios of a given sequence are always disjoint, and the events involved are independent. Then, the following condition is always valid, given we are in the probability space $\Omega^3$ of the sequence $\{a \rightarrow b \rightarrow c\}$:

$$\sum_{i=1}^{N} P(S_i) = 1,$$

where $N = 8$ represents the total number of scenarios associated with the sequence $\{a \rightarrow b \rightarrow c\}$.

In the analysis of accidents and operational safety, it is a matter of identifying and evaluating the likelihood of the sequences of events that may deviate the system from the safe operating domain. The occurrence of the feared events is random in time. This means that the intervals separating the successive occurrences of the events $(\Delta_1, \Delta_2, \Delta_3)$ (Figure 6.1) are random variables under a single constraint, which is $\Delta_1 + \Delta_2 + \Delta_3 \leq T$. The nature of the feared events is clearly identified, but the instants of occurrence are random. Operational safety and accident analysis use almost the same tools, methods, models and failure data, but from two different points of view. Operational safety looks at predicting the occurrence likelihood of an accident with time. It is a preventive standpoint of view. Accident analysis takes place after the occurrence of accidents. It provides knowledge about the evolution of the accident and its consequences. The lessons learnt from accident analysis provide the rationales to improve the models and the data used by the predictive models.

These feared events are called "initiating" events in certain industrial sectors such as in the nuclear safety sector. In order to characterize them and integrate them into an event tree, the feared events are the subject of in-depth analyses by fault trees. Because of fault tree analysis, the occurrence rates of feared events will be determined. These occurrence rates are the data required for determining the occurrence probabilities of the consequences that feared events imply.

In safety analysis, we are interested in two types of sequences: operational transient sequences, carried out by the system, and the sequences of actions undertaken by the operators. These two types of sequences are often mixed in the same scenario. The nature of the desired (or feared) events (system transients or operator actions) is well identified and the intervals separating the successive occurrences of the events $(\Delta_1, \Delta_2, \Delta_3)$

(Figure 6.1) are a priori fixed by the constraint, which is $\Delta_1 + \Delta_2 + \Delta_3 \leq T$. This means that the intervals separating successive occurrences of events $(\Delta_1, \Delta_2, \Delta_3)$ are prefixed, but the instant of occurrence of events in the prefixed intervals remains random.

So, two distinct approaches are identified in sequential events analysis:

– analysis with variable time intervals (dynamic analysis);

– analysis with fixed time intervals (semi-dynamic).

## 6.2. Sequential analysis with variable time intervals

Let $S$ be a sequence of events composed of $n$ independent events, which take place in this order $\{e_1 \rightarrow e_2 \rightarrow e_3 \rightarrow \cdots \rightarrow e_n\}$. The probability $P_n(T)$ of occurrence in a time interval $T$ of a sequence of $n$ independent events $e_i$, $(i = 1,2, \ldots, n)$ is determined by the following multiple integral:

$$P_n(T) = \int_0^T \rho_1(\xi_1)d\xi_1 \int_{\xi_1}^T \rho_2(\xi_2)d\xi_2 \int_{\xi_2}^T \rho_3(\xi_3)d\xi_3 \;\ldots\ldots\; \int_{\xi_{n-1}}^T \rho_n(\xi_n)d\xi_n, \qquad [6.1]$$

where $\rho_i(\xi)$ is the occurrence probability density of the event $i$ over time and $i$ indicates the order of occurrence.

Equation [6.1] can also be expressed in a more reduced form as:

$$P_n(T) = \int_0^T P_{n-1}(\xi).\rho_n(\xi_n)d\xi_n. \qquad [6.2]$$

The probability of occurrence $P_n(T)$ can have an analytical solution if the occurrences of events follow a stochastic Poisson process[1]. Otherwise, the integral in equation [6.1] can be solved by suitable numerical methods, among which we can cite the Monte Carlo simulation method.

1 Eid, M. (2011). A general analytical solution for the occurrence probability of a sequence of ordered events following Poison stochastic processes. *Journal of Reliability Theory & Applications*, RT&A # 03, 2(22), September. ISSN 1932-2321. (*Electronic journal of international group on reliability, registered in the library of the USA congress*).

In the case of a sequence with $n$ events whose occurrence follows a stochastic Poisson process, the integral in equation [6.2] will be rewritten as follows:

$$P_n(T) = \int_0^T P_{n-1}(\xi).e^{-\nu_n \xi}.\nu_n d\xi, \qquad [6.3]$$

where $\nu_n$ is the occurrence rate of the $n$th event and $P_0(T) = 1$.

The general solution of equation [6.3] takes the following form[2]:

$$P_n(t) = \sum_{j=0}^{n} C_j^n.e^{-\omega_j^n t}, \qquad [6.4a]$$

where:

$$\omega_0^1 = \nu_1, \omega_1^1 = 0, C_1^1 = -C_0^1 = 1,$$

$$\omega_j^n = \omega_j^{n-1} + \nu_n, \omega_n^n = 0, \forall j \in [0, n-1],$$

$$\omega_j^n C_j^n = -\nu_n C_j^{n-1} \ \forall j \in [0, n-1], C_{n+1}^{n+1} = -\left(\sum_{j=1}^{n} C_j^n\right), \ C_1^1 = 1.$$

Since $\omega_n^n$ is always zero, the probability of occurrence can be written as follows:

$$P_n(t) = \sum_{j=0}^{n-1} C_j^n.e^{-\omega_j^n t} + C_n^n. \qquad [6.4b]$$

For a sequence $S$ composed of four independent events $\{e_1 \to e_2 \to e_3 \to e_4\}$, the probability of occurrence $P_4(t)$ in an interval $t$ will therefore be determined by:

$$P_4(t) = C_0^4 e^{-\omega_0^4 t} + C_1^4 e^{-\omega_1^4 t} + C_2^4 e^{-\omega_2^4 t} + C_3^4 e^{-\omega_3^4 t} + C_4^4.$$

We are going to look at the evolution over time of the sequence $S$ and the reversed sequence $S^-$ (Figure 6.2), where the sequence $S$ is defined by $\{e_1 \to e_2 \to e_3 \to e_4\}$ and the reversed sequence $S^-$ is defined by $\{e_4 \to e_3 \to e_2 \to e_1\}$, with the same independent events $\{e_1, e_2, e_3, e_4\}$. The occurrence of the events follows a stochastic Poisson process, characterized by a constant occurrence rate $\nu$.

2 Ibid.

The occurrence rates of the events $\{e_1, e_2, e_3, e_4\}$ are given in Table 6.1.

| $i$ | 1 | 2 | 3 | 4 |
|---|---|---|---|---|
| $\nu_i(\mathbf{h}^{-1})$ | $8 \times 10^{-1}$ | $4 \times 10^{-1}$ | $2 \times 10^{-1}$ | $1 \times 10^{-1}$ |

**Table 6.1.** *Numerical values of the occurrence rates ($h^{-1}$)*

The occurrence probability $P_4(t)$ in an interval $t$ of the sequence $S$ is determined by equation [6.5a]:

$$P_4(t) = 3.17 \times 10^{-3} e^{-1.5t} - 4.76 \times 10^{-2} e^{-0.7t} + 2.22 \times 10^{-1} e^{-0.3t} - 3.81 \times 10^{-1} e^{-0.1t} + 2.03 \times 10^{-1}. \quad [6.5a]$$

For the reversed sequence $S^-$ composed of the same four independent events $\{e_4 \rightarrow e_3 \rightarrow e_2 \rightarrow e_1\}$, the probability of occurrence $P_4(t)$ in an interval $t$ is determined by:

$$P_S^-(t) = 2.03 \times 10^{-1} e^{-1.5t} - 3.81 \times 10^{-1} e^{-1.4t} + 2.22 \times 10^{-1} e^{-1.2t} - 4.76 \times 10^{-2} e^{-0.8t} + 3.17 \times 10^{-3}. \quad [6.5b]$$

The numerical values used in equations [6.5a] and [6.5b] are determined using the formulas detailed in Tables 6.2 and 6.3.

The occurrence probabilities of sequences $S$ and $S^-$ are plotted in Figure 6.2.

| $n$ | $\omega_0^n =$ | $\omega_1^n =$ | $\omega_2^n =$ | $\omega_3^n =$ | $\omega_4^n =$ |
|---|---|---|---|---|---|
| 1 | $\omega_0^1 = \nu_1$ | $\omega_1^1 = 0$ | | | |
| 2 | $\omega_0^2 = \omega_0^1 + \nu_2$ | $\omega_1^2 = \omega_1^1 + \nu_2$ | $\omega_2^2 = 0$ | | |
| 3 | $\omega_0^3 = \omega_0^2 + \nu_3$ | $\omega_1^3 = \omega_1^2 + \nu_3$ | $\omega_2^3 = \omega_2^2 + \nu_3$ | $\omega_3^3 = 0$ | |
| 4 | $\omega_0^4 = \omega_0^3 + \nu_4$ | $\omega_1^4 = \omega_1^3 + \nu_4$ | $\omega_2^4 = \omega_2^3 + \nu_4$ | $\omega_3^4 = \omega_3^3 + \nu_4$ | $\omega_4^4 = 0$ |

**Table 6.2.** *The coefficients $\omega_j^n$ for n = 1, 2, 3, 4 and j = 0, 1, 2, 3, 4*

| $n$ | $C_0^n =$ | $C_1^n =$ | $C_2^n =$ | $C_3^n =$ | $C_4^n =$ |
|---|---|---|---|---|---|
| 1 | $C_0^1 = -1$ | $C_1^1 = +1$ | | | |
| 2 | $C_0^2 = -\nu_2 C_0^1/\omega_0^2$ | $C_1^2 = -\nu_2 C_1^1/\omega_0^2$ | $C_2^2 = -\sum_{j=0}^{1} C_j^2$ | | |
| 3 | $C_0^3 = -\nu_3 C_0^2/\omega_0^3$ | $C_1^3 = -\nu_3 C_1^2/\omega_1^3$ | $C_2^3 = -\nu_3 C_2^2/\omega_2^3$ | $C_3^3 = -\sum_{j=0}^{2} C_j^3$ | |
| 4 | $C_0^4 = -\nu_4 C_0^3/\omega_0^4$ | $C_1^4 = -\nu_4 C_1^3/\omega_1^4$ | $C_2^4 = -\nu_4 C_2^3/\omega_2^4$ | $C_3^4 = -\nu_4 C_3^3/\omega_3^4$ | $C_4^4 = -\sum_{j=0}^{3} C_j^4$ |

**Table 6.3.** *The coefficients $C_j^n$ for n = 1, 2, 3, 4 and j = 0, 1, 2, 3, 4*

The sequences $S$ and $S^-$ themselves do not follow a stochastic Poisson process as their basic events. However, they follow a stationary stochastic process, characterized by an asymptote. The occurrence probability of the process $S$ reaches the asymptotic value of $2 \times 10^{-1}$ and that of the sequence $S^-$reaches the asymptotic value of $3.2 \times 10^{-3}$ (Figure 6.2).

Let us not forget that the sequence $S\{e_1 \to e_2 \to e_3 \to e_4\}$ is just a sequence among the 16 sequences ($2^4$) that form the probability space to which this sequence belongs. If the occurrence of the 16 sequences were uniformly probable, each sequence would have had a stationary (asymptotic) probability equal to $\{\frac{1}{16} = 6.25 \times 10^{-2}\}$. But the 16 sequences are not. The stationary probability $P_\infty$ of the sequence $S$ is rather equal to $2 \times 10^{-1}$ (Figure 6.2).

Even if the stochastic process governing the sequence $S$ is not a Poisson process, we can still characterize its occurrence by its asymptotic behavior, its occurrence rate as a function of time, the most probable interval for the completion of the whole sequence and the average time interval for the completion of the sequence.

We are therefore going to determine the probability density of occurrence $D_n(t)$ of a sequence with $n$ independent by:

$$D_n(t) = \frac{d}{dt}P_n(t) = \sum_{j=0}^{n} C_j^n . \omega_j^n e^{-\omega_j^n t}. \qquad [6.6]$$

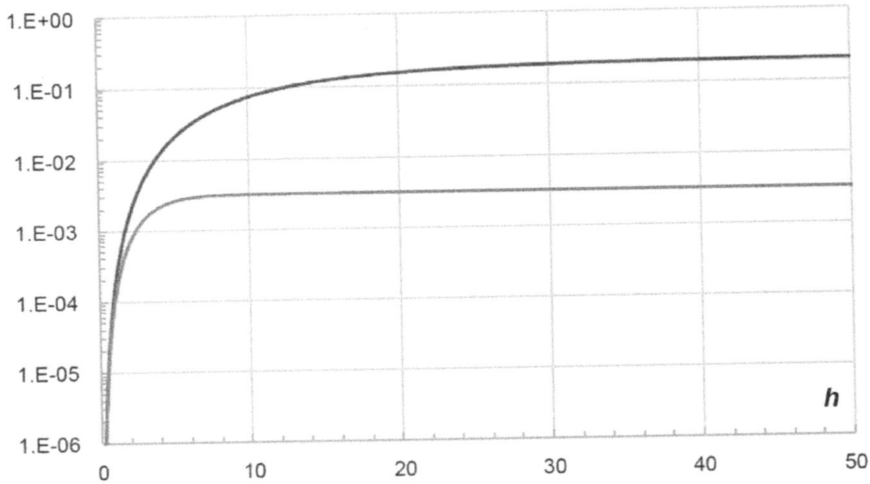


**Figure 6.2.** *The evolution over time of the probability $P_n(t)$ of the sequence S (blue) and of the reversed sequence $S^-$ (red). For a color version of this figure, see www.iste.co.uk/elhami/reliability.zip*

Not surprisingly, this process dies out with time, $\lim_{t\to\infty} D_n(t) \to 0$ (equation [6.6]). This demonstrates the stationary nature of the process, which is explained by the fact that all the events involved in the sequence follow a stochastic Poisson process.

With regard to the sequences $S$ and $S^-$, the probability densities $D_4(t)$ are plotted in Figure 6.3 as a function of time. The most probable time interval that these sequences can occupy is of the order of 8 h for the sequence $S$ and 2.5 h for the sequence $S^-$, as we can see in Figure 6.3.

The behavior over time of the probability densities plotted in Figure 6.3 is a characteristic behavior of all sequences when the basic events follow stochastic Poisson processes.

We can also determine the average time $\bar{T}$ for the realization of the sequence:

$$\bar{T} = \lim_{T\to\infty} \left[ \left( \int_0^T t dP_n(t) \right) / \left( \int_0^T dP_n(t) \right) \right] = -\frac{1}{P_\infty} \cdot \sum_{j=0}^{n-1} \frac{c_j^n}{\omega_j^n}. \qquad [6.7]$$

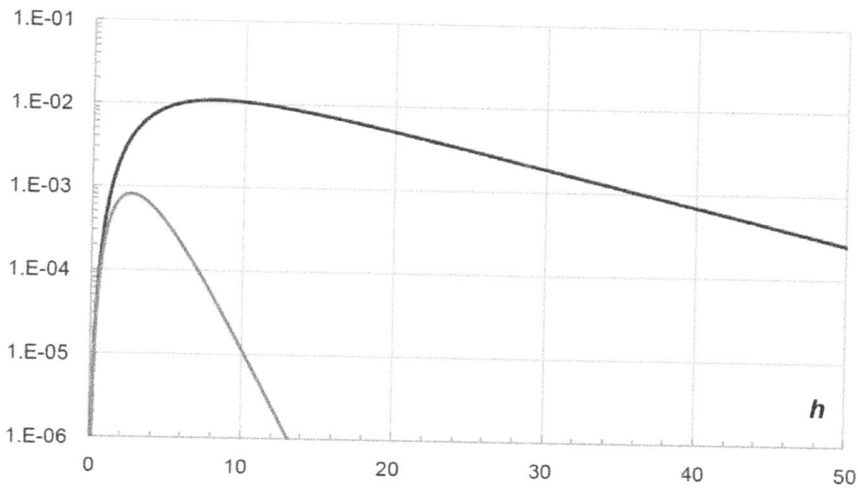


**Figure 6.3.** *The probability densities $D_n(t)$ of the sequences $S$(blue) and $S^-$(red). For a color version of this figure, see www.iste.co.uk/elhami/reliability.zip*

If we consider the sequences $S$ and $S^-$ previously discussed, their average completion times are of the order of ~15.5 h and 3.5 h, respectively.

We can also determine the occurrence rate of the sequence $\Lambda_n(t)$ as:

$$\Lambda_n(t) = \frac{D_n(t)}{P_n(t)}. \qquad [6.8]$$

It is important to remember that the time axis of Figures 6.2, 6.3 and 6.4 does not describe an instant, but the length of a time interval. The rate of occurrence (in Figure 6.4) is plotted against the time interval of the execution of the sequence. These time intervals are all measured from the same origin $t_0$, which is the occurrence of an initiating event "initiator".

In addition, the order of occurrence of the basic events impacts the dynamic characteristics of the sequence. This is not the case with static models of sequential calculation such as the model with fixed intervals.

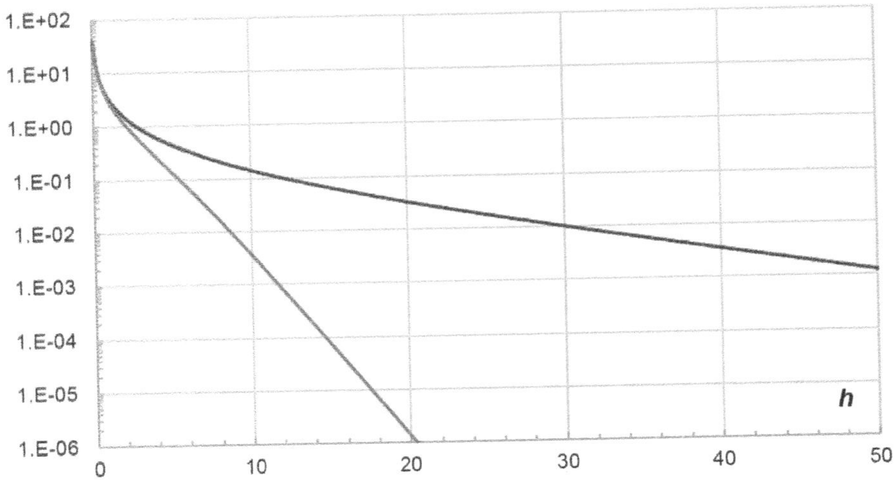


**Figure 6.4.** *Occurrence rate $\Lambda(t)(h^{-1})$ of sequences $S$(blue) and $S^-$(red). For a color version of this figure, see www.iste.co.uk/elhami/reliability.zip*

## 6.3. Sequential analysis with fixed time intervals

Historically, the most common practices in safety analysis tend to favor static models because of their conservatism. The occurrence of each event $e_i$ will be fixed in a time interval $\Delta_i$ in which it must occur if it occurs. Any occurrence outside of this prefixed time window is not considered. Equation [6.1] is no longer a multiple integral, but a multiplication of integrals like:

$$P_n(T) = \prod_{j=1}^{n} \int_{t_{j-1}}^{t_j+\Delta_j} \rho_j(\xi_j)d\xi_j = \prod_{j=1}^{n} h_j(\Delta_j), \quad [6.9]$$

where the intervals of occurrence $\Delta_j$ are well defined, such that:

$$t_j = t_{j-1} + \Delta_j,\ t_0 = 0,\ j = 1, 2, \dots, n, \text{ and } \sum_{j=1}^{K} \Delta_j = T.$$

$h_j(\Delta_j)$ is the occurrence probability of the event "$e_j$" within the time interval $\Delta_j$ and is determined by:

$$h_i(\Delta_i) = \int_{t_{i-1}}^{t_i=t_{i-1}+\Delta_i} \rho_i(\xi)d\xi. \quad [6.10]$$

If the occurrence of each event "$i$" follows a stochastic Poisson process characterized by a constant occurrence rate "$\nu_i$", the occurrence probability $h_i(\Delta_i)$ of the event "$e_i$" in the time interval $[\Delta_i]$ is determined by:

$$h_i(\Delta_i) = [1 - \exp(-\nu_i \Delta_i)]. \quad [6.11]$$

Note that the probability $g_i(\Delta_i)$ of the non-occurrence of the event "$e_i$" in the time interval $[\Delta_i]$, $(\forall i;\ i = 1,2, \dots)$ is determined by:

$$g_i(\Delta_i) = \exp(-\nu_i \Delta_i). \quad [6.12]$$

Let $S\{e_1 \rightarrow e_2 \rightarrow e_3 \rightarrow e_4\}$ be a sequence made up of four independent events. We wish to determine the occurrence probability $P_4(T = 20)$ of this sequence, given that the time windows of occurrence of these events are all equal to 5 h, $\{5, 5, 5, 5\}$. The occurrence rates and the imposed periods are given in Table 6.3. Note that this is the same sequence $S$ as the one treated previously.

How about the probability of occurrence of the reverse sequence $S^{-1}$?

| $i$ | Unit | 1 | 2 | 3 | 4 | Total |
|---|---|---|---|---|---|---|
| $\nu_i$ | $h^{-1}$ | $8 \times 10^{-1}$ | $4 \times 10^{-1}$ | $2 \times 10^{-1}$ | $1 \times 10^{-1}$ | – |
| $\Delta_i$ | h | 5 | 5 | 5 | 5 | – |
| $h_i$ (5) | – | $9.8{,}168^{-1}$ | $8.6{,}466^{-1}$ | $6.3{,}212^{-1}$ | $3.9{,}347^{-1}$ | – |
| $P_4(20)$ | – | – | – | – | – | $2.11^{-1}$ |

**Table 6.4.** *Occurrence probabilities of events and sequence S*

The comparison between the dynamic and the semi-dynamic sequential calculations is summarized in Table 6.5.

| Sequence | Dynamic model, see Figure 6.2 (20 h) | Static model (4 × 5 h) |
|---|---|---|
| Sequence ($S$) | $1.5 \times 10^{-1}$ | $2.11^{-1}$ |
| Sequence ($S^{-1}$) | $3.3 \times 10^{-3}$ | $2.11^{-1}$ |

**Table 6.5.** *Comparison between occurrence probabilities*

As expected, the semi-dynamic model does not take into account the order of occurrence of the events in the sequence. In Table 6.5, we find the same probability of occurrence for the sequence $S$ as for the reversed sequence $S^{-1}$.

## 6.4. Conclusion

To conclude on the modeling of the sequential events, one may indicate that the sequential calculation by semi-dynamic model is conservative because:

1) it overestimates the occurrence probability of occurrence compared to the dynamic sequential calculation at least from the safety standpoint of view;

2) it quite often produces values close to the asymptotic values of the occurrence probability of the sequence. This is because the choice of the time windows of events' occurrence is often motivated by safety reasons.

Static models are also easier to compute, and they propagate the uncertainties that may exist in the basic events occurrence data.

It should also be noted that, historically, conservatism is a reassuring value in terms of operational safety.

For all these reasons, static models are widely practiced in the field of system safety assessment.

The same would not be said for accident analysis, where conservatism or an overestimation of the occurrence of accidents could lead to unacceptable consequences in terms of economic and technological costs, while jeopardizing the credibility of decision-making institutions.

# 7

# The Monte Carlo Simulation Method

The Monte Carlo simulation method is widely used in system reliability and safety analysis. It reveals its full power with the growth of the complexity of the systems. The complexity of a system is not limited to its size, to the number of its constituents and subsystems, to the extent of the physical phenomena in action or to the heterogeneity of the stochastic processes involved, but it also integrates the structural, organizational and regulatory aspects that frame the missions and functionalities of the system. The Monte Carlo method makes it possible to go beyond the limits of the analytical formalism.

We do not intend to cover the Monte Carlo method exhaustively: its theoretical bases and the extent of its fields of application. This is beyond the restricted scope of the book. As soon as the basic notions are briefly presented, we will move on to applications with the same care given to the didactic condition as in the previous chapters.

## 7.1. Generation of random variables

Simulation by the Monte Carlo method is based on its ability to generate random variables following a well-defined probabilistic distribution. For example, if we are interested in the uptime before failure of a given component and if we know the probabilistic law that governs the occurrence of its failures, we can then generate this random variable called "Time To Fail – TTF". We proceed exactly the same way regarding the repair time "Time To Repair – TTR".

Putting the Monte Carlo method into practice requires the use of random variables generators. Two approaches are, in principle, possible to generate random variables: analogical and numerical. We are more interested in the numerical generation of random variables.

Some functions called pseudo-random allow us to develop basic generators of random variables with specific probabilistic characteristics. But are we going to develop random variable generators for each family of probabilistic laws?

Certainly not! It suffices to use basic generators of random variables to generate random variables belonging to other and more complex probabilistic laws. We precisely have generators of (pseudo)-random variables (almost) continuous and uniform in the interval ]0,1[.

Currently, every *software* conceived for numerical calculations has its basic generator of random variables under different names such as Alea (), Ran() and Random(). This is particularly the case for applications such as Excel, Mathlab, Scilab or Mathematica, and almost all compilers are equipped with basic generators. They are even integrated into the most elementary public calculators and mobiles.

This basic numerical brick is the starting point for any numerical simulation, regardless of the complexity of the simulation case. We always start with the generation of sequences of real numbers that can statistically be considered as sequences of random variables uniformly distributed between 0 and 1.

The linear congruent generator (LCG) is the most widely used of these basic generators. These generators produce a sequence of numbers $(X_n)$ defined by the following recurrence equation:

$$\xi_{n+1} = (a\xi_n + c) \; mod \; m,$$

where $a$, $c$, $m$ and $\xi_0$ are positive integers called multiplier, increment, modulo and seed, respectively. The sequence $(\xi_n)$ is periodic at an interval equal to the value of the modulo $m$. In order to obtain a sufficiently long sequence of random numbers (a long period), it is necessary to have a sufficiently large modulo $m$. For example, for a computer with 32-bit addressing structure, one of the generators used by IBM is:

$$\xi_{n+1} = (7^5 \times \xi_n - 1) \; mod \; 2^{32}.$$

It is obvious that the variables $\xi_{n+1}$ are indeed deterministic. Each time we use the same grain $\xi_0$, we get exactly the same sequence $\{\xi_n; n = 0,1,...\}$. But the variables behave statistically as if they were random variables that follow a uniform distribution between 0 and 1.

This pseudo-random variable will be designated throughout this chapter by the Greek letter $\xi$, and it belongs to the open interval $]0,1[$.

## 7.2. Sampling

Random variables are associated with everyday life and have a varying nature such as the arrival time of a bus, the amount of rains that fall on a given surface on a given date, the wind speed in a place on a given date, the height of a sea wave, the effective expiry date of a pot of yoghurt, the life expectancy of a living being (plant, animal, human) and the time interval between two radioactive decays in a radioactive material. Rare are the cases that can be modeled directly using random variables uniformly distributed in the interval $]0,1[$.

We must therefore manage to build stochastic models describing natural phenomena by using the basic generators of random variables[1] in the interval $]0,1[$. Sampling is therefore a process that transforms a basic random variable coming directly from a basic random variable generator into a random variable associated with a real event of interest. Sampling can be natural or accelerated in different ways.

We are going to treat only the sampling without acceleration and more precisely the analytical bijective sampling without acceleration.

## 7.3. Analytical bijective sampling

Let $C$ represent the operating state and $\bar{C}$ the failure state of an elementary component well defined by its characteristic parameters $(\lambda, \mu, \gamma)$, where the parameters $(\lambda, \mu)$ are constant over time. The set of random variables $\tau_i$ $\{\tau_i; i = 1,2, ..., n\}$ represents the simulated "times To fail, TTF".

1 Pseudo-random.

As we have already presented in Figure 1.1, the TTF is a random variable. The operation histogram (in Figure 1.1) was generated by simulation using the Monte Carlo method. We are interested in the development of a model to generate this numerical random variable in order to numerically simulate the failure of the component $C$. We need to perform a numerical sampling of the component's TTF. A sampling campaign is made up of several random failure histories (/samples).

How can we pass from the basic random variable $\xi$ ($\xi;\ \xi \in ]0,1[$) to the random variable $\tau$ ($\tau;\ \tau \in [0,\infty[$)?

Let $\rho(t)$ be the probability density function that governs the occurrence of the failure of the investigated component defined by:

$$\rho(t) = \lambda e^{\lambda t}. \quad [7.1]$$

Since the basic random variable $\xi$ is uniform in the interval ]0,1[, its probability density function $h(\xi)$ is as follows:

$$h(\xi) = 1.$$

If, and only if, a bijective relationship exists between the two random variables $\xi$ and $t$, that is, each point $\xi$ corresponds to a single point $t$ and vice-versa, "the segment $[0, t]$ is the image of the segment $[0, \xi]$ and vice versa" (Figure 7.1).

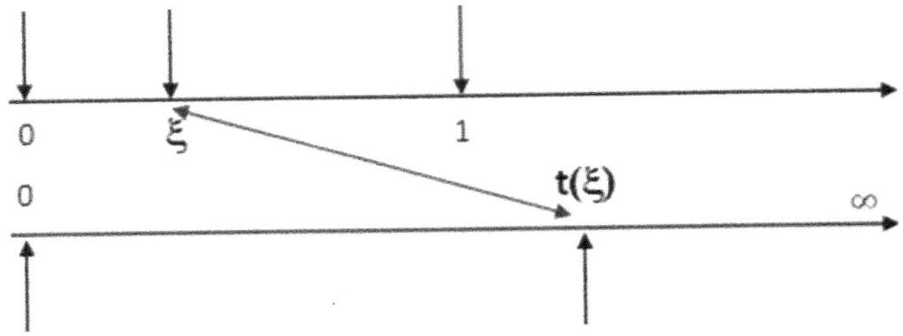


**Figure 7.1.** *Schematic representation of the bijective mapping between $\xi$ and $t$*

We can therefore state that "the probability that the random variable $t'$ exists in the interval $[0, t]$ is necessarily equal to the probability that the random variable $\xi'$ exists in the interval $[0, \xi]$" such as:

$$\int_0^t \rho(t')dt' = \int_0^\xi h(\xi')\, d\xi',$$

submitting each probability density function in the integral allows us to write:

$$\int_0^t (\lambda e^{-\lambda t'})dt' = \int_0^\xi (1)\, d\xi'.$$

Thus:

$$\left[1 - e^{-\lambda t}\right] = \xi.$$

A bijective relationship then exists between the two random variables such as:

$$t_f = \frac{1}{\lambda}\ln\left(\frac{1}{1-\xi}\right). \qquad [7.2]$$

Equation 7.2 is an analytical bijective mapping function between two random variables. Obtaining an analytical bijective relation is not always possible. In this example, analytical sampling was possible because the probability density $\rho(t)$ is integrable and the inverse analytical function of its integral exists. If any of the two preceding conditions is missing analytical bijective sampling will not be possible. In that case, numerical sampling exploiting tables either of the probability density, of the integral of the probability density or of the inverse function of the integral will be the only option to carry on the simulation. This is the case, for example, if the probability density is in the form of a Gaussian distribution. The integral of a Gaussian distribution does not have an analytical solution but a tabulated one.

We designate the time to fail (TTF) in equation [7.2] by $t_f$. If we are interested in the time to repair $t_r$ (TTR), we can apply the same process and find that:

$$t_r = \frac{1}{\mu}\ln\left(\frac{1}{1-\xi}\right). \qquad [7.3]$$

We therefore conclude that for any component well defined by its characteristic parameters, we can numerically simulate its operation histogram $(t_f, t_r)$ (Figure 1.1).

So, how can we simulate the reliability and the availability of a system from the operation of its independent components?

## 7.4. Generic Monte Carlo simulation approach

A Monte Carlo simulation is constructed according to the algorithmic procedure described in Algorithm 7.1.

1) Define the number of sampling batches $N$ to be carried out and the number of histories $L$ per batch.

2) Choose the mathematical formulation that simulates the studied phenomenon and associate an adequate characteristic score (estimator) to each history $l$, $(l;\ l \in [1, L])$.

3) For the $n^{th}$ batch, $(n; n \in [1, N])$, carry out $L$ simulation histories of the studied phenomenon and store the independent $L$ histories. For each completed history, collect the characteristic scores $s_l$, $(s_l;\ l = 1, 2, \ldots, L)$. Then, estimate the mean score $S_n$ of the $n$th batch:

$$S_n = \frac{1}{L}\left(\sum_{l=1}^{L} s_l\right).$$

4) Repeat the previous step #3 until we sample all the $N$ batches and their mean scores $S_n$ , $(S_n, n = 1, 2, 3, \ldots, N)$.

5) Determine the mean score $E_N$ (the estimator) of the mean scores $S_n$ and its variance $\sigma_N^2$:

$$E_N = \frac{1}{N}\sum_{n=1}^{N} S_n,$$

$$E_N^2 = \frac{1}{N}\sum_{n=1}^{N} (S_n)^2,$$

$$\sigma_N^2 = \frac{1}{N}\left[E_N^2 - (E_N)^2\right].$$

6) If $N$ and $L$ are large enough ($N \to \infty$, $L \to \infty$), we can state that the distribution of the batches mean score $S_n$ converges to a Gaussian distribution, whose expectation ($\mu$) and variance ($\sigma^2$) are the convergent values of $S_n$ and $\sigma_N^2$ such as:

$$\lim_{\substack{L\to\infty \\ N\to\infty}} S_n \to \mu \qquad \text{(the law of large numbers),}$$

$$\lim_{\substack{L\to\infty \\ N\to\infty}} \sigma_N^2 \to \sigma^2 \qquad \text{(the central limit theorem).}$$

7) Given that $N$ and $L$ are large enough ($N \to \infty$, $L \to \infty$), we may consider that the estimator $E_N$ is close enough to the mathematical expectation $\mu$ characterizing a Gaussian distribution whose variance is $\sigma^2 \approx \sigma_N^2$.

8) END.

9) STOP.

**Algorithm 7.1.** *Monte Carlo simulation generic algorithm*

It is important to underline the conditions of the application of the law of large numbers (weak convergence) and the central limit theorem (strong convergence). Both constitute the theoretical foundation of the Mont-Carlo simulation. Neither the first nor the second poses quantitative criteria that allow the affirmation that $N$ and $L$ are large enough, ($N \to \infty$, $L \to \infty$). In practice, it depends exclusively on the nature of the treated problem. For some cases, a sample of a small population (~10–50 histories) would suffice, and for others a few million samples may be needed.

We can generally state that "one may affirm that $L$ is large enough ($L \to \infty$) if:

$$\frac{\sum_{l=1}^{L2} s_l}{L2} \approx \frac{\sum_{l=1}^{L1} s_l}{L1}, \forall \quad L2 > L1,$$

where $L1$ and $L2$ are the number of histories simulated in two different simulation batches of the same case.

That means, when the estimator no longer varies significantly with the sample size $L$, that the estimator is tending to converge.

As soon as a large enough value $L_\infty$ is found, the large enough value $N_\infty$ will be the one corresponding to an acceptable variance $\sigma_N^2$ (or standard deviation $\sigma_N$):

$$\sigma_N^2 = \frac{1}{N}\left[E_N^2 - (E_N)^2\right].$$

The acceptability can also be determined based on the ratio of the standard deviation to the estimator:

$$\varepsilon = \frac{\sigma_N}{\mu}.$$

In some applications, a $\varepsilon$ ratio of 15%–10% might be acceptable, and in others the acceptable figure may be required to be within the range of $(10^{-5} - 10^{-6})$. We underline thus that the standard deviation is inversely proportional to $\sim \sqrt{N}$.

A practical conclusion is that the sufficiency of the samples size ($L$) is judged by the convergence of the estimator. However, the sufficiency of the batches number ($N$) is judged by the convergence of the overall variance $\sigma_N^2$ toward a minimal value.

As soon as the conditions of a large sample $(N \to \infty, L \to \infty)$ are satisfied, the distribution of the estimated mean score $E_N$ (the estimator) approaches a Gaussian distribution, whatever is the real distribution of the basic phenomena, according to the central limit theorem.

Having established an overview of the generic Monte Carlo simulation demarch, let us proceed for applications.

## 7.5. Simulation of transitions in a multistate system

Consider the case of a system presented by its state graph (Figure 7.2). The system represents a crisis management model. The crisis management model is schematically represented by a stochastic process with four possible states. This is the simplest representation that one can imagine for a crisis management model.

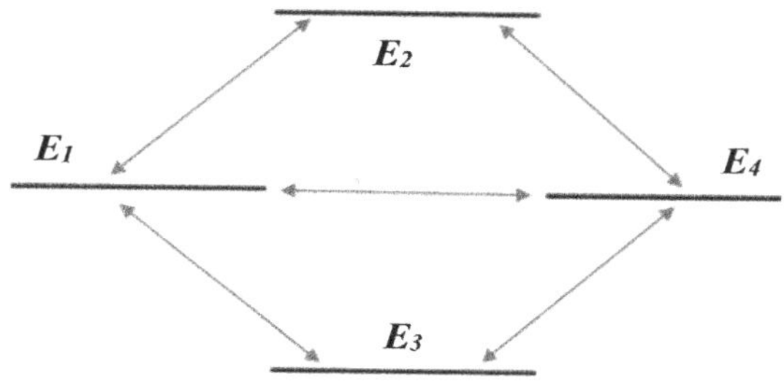


**Figure 7.2.** *State graph representing a multistate system. For a color version of this figure, see www.iste.co.uk/elhami/reliability.zip*

The states are defined as follows:

– $E1$ is the safe state, or the "cold phase", which represents the operational normality;

– $E2$ is the state of crisis identification, assessment and decision-making by the Crisis Management Group, including data collection and modeling, analysis, simulation, proposal of measures and monitoring of their implementation by the field intervention operators;

– $E3$ is the state of observation and monitoring of the crisis evolution on the ground, in reaction to the decisions taken in $E2$: implementations, monitoring the crisis evolution tendency on the ground and field data updating feedback;

– $E4$ is the state of "active crisis" or the "hot phase".

The criteria of crisis management failure are defined as follows:

– sojourn time longer than 4 h (240 min) in the state $E4$, at least once;

– cumulated sojourn time longer than 4 h (240 min) in the states $E2$ and $E3$.

The correspondent transition matrix is given in Table 7.1. The elements of the transition matrix are the transition rates $\tau_{ij}$ from state $i$ to state $j$. Notice that the transition rates $\{\tau_{11}, \tau_{22}, \tau_{33}, \tau_{44}, \tau_{23}, \tau_{32}\}$ are null. Thus, we propose to set the rates $\{\tau_{11}, \tau_{22}, \tau_{33}, \tau_{44}, \tau_{23}, \tau_{32}\}$ to a very small value such as $10^{-20}$, when performing the numerical application. The transition rates are expressed in $\min^{-1}$.

As for the initial conditions, the system starts in the state $E1$ at $t = 0$.

| | $j = 1$ | $j = 2$ | $j = 3$ | $j = 4$ |
|---|---|---|---|---|
| $i = 1$ | 0.00E–00 | 4.00E–03 | 4.00E–03 | 1.00E–03 |
| $i = 2$ | 4.00E–03 | 0.00E–00 | 0.00E–00 | 1.00E–03 |
| $i = 3$ | 4.00E–03 | 0.00E–00 | 0.00E–00 | 1.00E–03 |
| $i = 4$ | 1.00E–03 | 1.00E–03 | 1.00E–03 | 0.00E–00 |

**Table 7.1.** *The transition matrix*

### 7.5.1. *The transition simulation approach*

For the simulation of the transitions in the case of the system described by the state graph in Figure 7.2, we will realize $N$ batches of $10^2$ transitions ($L = 10^3$) each. We call a numerical simulation of $N$ batches with $L$ stories per batch a campaign.

A significant amount of information can be extracted from the simulation of the transitions between states, such as the sojourn times in each states, the number of transitions per state, the occurrence times of the first transition to the state $E4$, the times between two successive occurrences of transitions to the state $E4$ and the probability of failure of the management of the crisis (see failure criterion above).

We are going to focus on the analysis of the sojourn times in the states to deduce the probability of failure in the management of the crisis. The probability of failure is an indicator of the effectiveness of the crisis management process.

Several algorithms can be proposed for this Monte Carlo simulation campaign. We will propose the simplest and the most direct algorithm: an unbiased natural simulation, without acceleration, and with bijective analytical sampling. We take advantage of the fact that the values of the corresponding transition rates (Table 7.1) do not suggest the presence of one or more significantly rare transitions. The presence of some of the events whose occurrences are relatively rare compared to other events adds some complexity to the simulation. It would require the use of acceleration techniques to improve the statistical quality of the simulation output. The choice of a simple algorithm is in line with the didactic objectives of the book.

We therefore propose the following algorithm by focusing on the histogram of the transitions, the collection of the sojourn times and the number of sojourns per states.

### 7.5.2. Transitions simulation algorithm

To simulate transitions between the states, starting from state $E1$, we design the following algorithm:

1) Initialize the variables:

– the number of the transition: $n = 0$

– the "cumulative sojourn times, $TS(i)$" and the "number of sojourns, $No(i)$" in different states $i$:

$TS(i) = No(i) = 0, \qquad \forall\, i = 1, 2, 3, 4$

– the starting state for the $n^{th}$ transition:

$Ed(n) = 0$ // (departure state)

– the arrival state:

$Ea(n) = 0$ // (arrival state)

– Read ( $\tau_{ij}$, ( $i = 1, 2, 3, 4$), ( $j = 1, 2, 3, 4$)), $N\max$ .

2) $n = 1$, $Ed(1) = 1$.

3) Generate four random variables; ($\xi_1, \xi_2, \xi_3, \xi_4$); uniformly distributed in the interval ]0,1[.

4) Determine four sojourn times in the state $i$ corresponding to the four possible transitions:

$$\underset{i \neq j}{\Delta_{ij}} = \frac{1}{\tau_{ij}} Ln\left(\frac{1}{1-\xi_j}\right)$$

// $\Delta_{ii}(\tau_{ii} = 0) \rightarrow \infty$ (choose a very large number such as $10^{20}$ $mn$)

5) Determine the minimum sojourn time. : $\Delta' = \min(\Delta_{i1}, \Delta_{i2}, \Delta_{i3}, \Delta_{i4})$.

6) Identify the arrival state $Ea(n) = j_{\min}$, //where $j_{\min}$ is the arrival state identifier, which corresponds to the minimum sojourn time $\Delta'$.

7) $TS(i_{\min}) = TS(i_{\min}) + \Delta'$, $No(j_{\min}) = No(j_{\min}) + 1$.

8) If $n \geq Nmax$. GoTo 9

Otherwise: $n = n + 1$

$Ed(n) = Ea(n-1)$

GoTo 3

9) Statistical analyses and calculations of the sojourn probabilities per state, the probabilities of failure using the failure criterion, etc.

10) END.

11) STOP

**Algorithm 7.2.** *Transition simulation algorithm*

Steps 1–11 are generic for all algorithms that generate a histogram of transitions. Only the content of step 9 can vary from one application to another depending on the quantities to be calculated, the targeted precision ($\sigma/\mu$), the acceleration techniques used to minimize the uncertainties and reduce the computational time.

### 7.5.3. *Transition histogram simulation results*

The simulation of the transitions between the states in the case of the system described by the state graph in Figure 7.2 was carried out using the numerical data in Table 7.1. As mentioned previously, we are primarily interested in producing the transitions histogram representing the crisis management process evolution.

The simulated transitions within the first $10^4$ min (~7 days) of sampling are illustrated in Figure 7.3. This represents the first 77 transitions out of $10^3$ transitions. The $10^3$ transitions represent $1.42 \times 10^5$ min of system operational time. We have scored 129 cases of failure (sojourn time in the hot phase is beyond the allowed time criterion). We could distinguish 70 long sojourns in state $E2$, 55 long sojourns in $E3$ and four long sojourns in $E4$. It is obvious that the long sojourns in the states $E2$ and $E3$ are the major causes for the failures of the crisis management process.

To illustrate numerically this evolution pattern, we report the first 20 simulated transitions as described by the algorithm in Table 7.2. The numerical tracing of the simulated transition sequences should enable an appreciation of the simplicity of the algorithm.

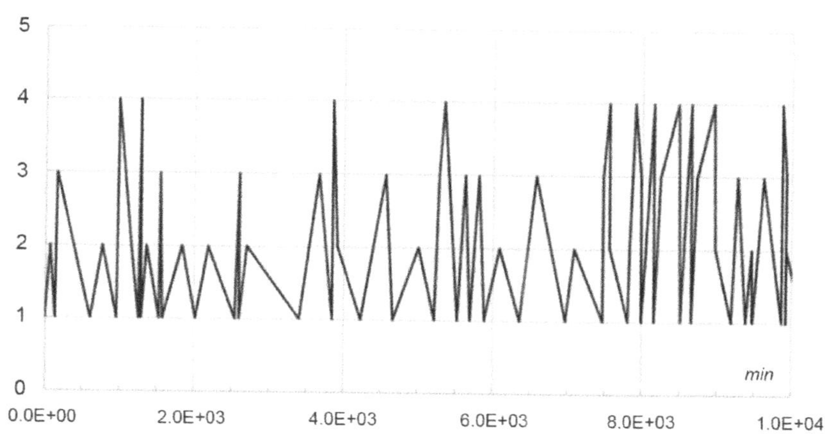


**Figure 7.3.** *The plot of the first 1E+4 min (77 transitions). For a color version of this figure, see www.iste.co.uk/elhami/reliability.zip*

| # | *I* | *T*1 | *T*2 | *T*3 | *T*4 | *T*_sojourn. (min) | *T*_Tot . (min) | *J* |
|---|---|---|---|---|---|---|---|---|
| | | | | | | | 0 | 1 |
| **1** | 1 | 8.62E+15 | 7.04E+01 | 5.09E+02 | 7.10E+02 | 7.04E+01 | 7.04E+01 | 2 |
| **2** | 2 | 5.92E+01 | 9.67E+14 | 2.09E+16 | 3.13E+03 | 5.92E+01 | 1.30E+02 | 1 |
| **3** | 1 | 7.74E+15 | 2.88E+02 | 2.59E+01 | 1.19E+03 | 2.59E+01 | 1.56E+02 | 3 |
| **4** | 3 | 4.53E+02 | 1.59E+16 | 1.91E+15 | 3.13E+03 | 4.53E+02 | 6.08E+02 | 1 |
| **5** | 1 | 2.24E+15 | 1.56E+02 | 1.66E+02 | 7.92E+03 | 1.56E+02 | 7.65E+02 | 2 |
| **6** | 2 | 1.93E+02 | 1.18E+16 | 4.81E+15 | 2.07E+03 | 1.93E+02 | 9.58E+02 | 1 |
| **7** | 1 | 3.74E+15 | 9.31E+02 | 5.25E+02 | 2.91E+01 | 2.91E+01 | 9.87E+02 | 4 |
| **8** | 4 | 2.38E+02 | 1.66E+02 | 4.41E+02 | 4.96E+15 | 1.66E+02 | 1.15E+03 | 2 |
| **9** | 2 | 9.64E+01 | 3.56E+16 | 1.40E+16 | 1.51E+03 | 9.64E+01 | 1.25E+03 | 1 |
| **10** | 1 | 1.28E+16 | 2.58E+02 | 1.08E+02 | 3.23E+01 | 3.23E+01 | 1.28E+03 | 4 |
| **11** | 4 | 5.44E+00 | 1.31E+03 | 3.69E+02 | 2.06E+15 | 5.44E+00 | 1.29E+03 | 1 |
| **12** | 1 | 1.22E+16 | 6.57E+01 | 1.97E+02 | 6.03E+02 | 6.57E+01 | 1.35E+03 | 2 |
| **13** | 2 | 1.73E+02 | 1.24E+16 | 6.46E+15 | 2.90E+02 | 1.73E+02 | 1.53E+03 | 1 |
| **14** | 1 | 4.59E+15 | 2.65E+02 | 9.19E+00 | 1.71E+02 | 9.19E+00 | 1.54E+03 | 3 |
| **15** | 3 | 3.90E+01 | 1.57E+16 | 7.23E+15 | 4.08E+02 | 3.90E+01 | 1.57E+03 | 1 |
| **16** | 1 | 1.04E+16 | 2.63E+02 | 1.72E+03 | 6.50E+02 | 2.63E+02 | 1.84E+03 | 2 |
| **17** | 2 | 1.78E+02 | 4.10E+15 | 1.13E+16 | 2.68E+03 | 1.78E+02 | 2.01E+03 | 1 |
| **18** | 1 | 5.24E+14 | 1.68E+02 | 2.03E+02 | 3.07E+03 | 1.68E+02 | 2.18E+03 | 2 |
| **19** | 2 | 3.66E+02 | 7.43E+15 | 6.77E+15 | 4.33E+02 | 3.66E+02 | 2.55E+03 | 1 |
| **20** | 1 | 1.96E+16 | 3.18E+02 | 4.90E+01 | 1.79E+03 | 4.90E+01 | 2.60E+03 | 3 |
| **21** | i | $T_{i1}^*$ | $T_{i2}^*$ | $T_{i3}^*$ | $T_{i4}^*$ | $T = \min(T1,T2,T3,T4)$ | $T(21) = T(20)+T$ | $j^{**}$ |

$^* T_{ij} = \frac{1}{\tau_{ij}} ln(\frac{1}{1-alea()})$.

$^{**}$ *I* and *J* correspond to the departure and the arrival states, respectively.

**Table 7.2.** *The first 20 transitions of the total 1,000 transitions simulation campaign*

### 7.5.4. *The number of sojourns by state*

We simulated $10^2$ batches of $10^3$ transitions each. The statistical treatment of the output data concerns the frequencies of sojourn by state. It is summarized in Table 7.3.

We notice that most of the sojourns are observed in the state $E1$ at 41.9%, followed by $E2$ and $E3$ with almost equal shares (23.2% and 22.7%, respectively), and finally $E4$ with a share equal to 12.2%. The estimated ratios of variance/mean are less than 3.2%. This represents a very acceptable uncertainty level.

Of course, we could have reduced further the uncertainties (the variance) if the number of batches $N$ went from $10^2$ to $10^3$. This would have divided the variance by $\sim\sqrt{10}$. But no significant interest can be expected, as long as the average number of sojourn is almost stationary, and the level of uncertainty is acceptable.

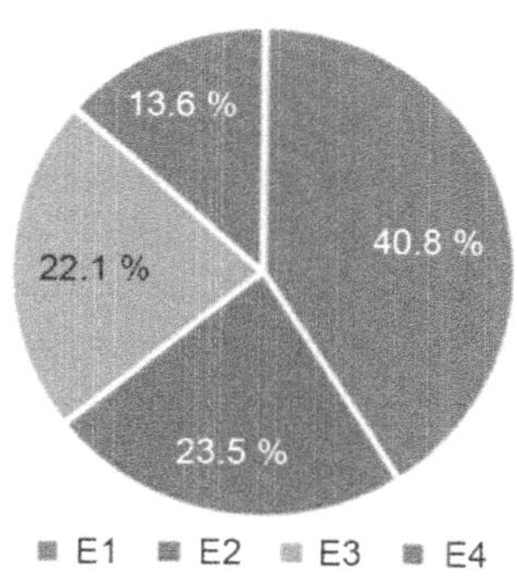


**Figure 7.4.** *Breakdown of average number of sojourn by state. For a color version of this figure, see www.iste.co.uk/elhami/reliability.zip*

| | $E1$ | $E2$ | $E3$ | $E4$ | |
|---|---|---|---|---|---|
| $(\sigma/\mu)$ % | 3.2% | 3.2% | 3.2% | 3.1% | |
| $\sigma$ | 1.32E−02 | 7.32E−03 | 7.16E−03 | 3.84E−03 | |
| $\mu \approx E(No)$ | 4.19E−01 | 2.32E−01 | 2.27E−01 | 1.22E−01 | |
| $No$ | 419 | 232 | 227 | 122 | 1,000 |

**Table 7.3.** *Average number of sojourns per state*

### 7.5.5. *Sojourn time by state*

After having studied the frequency of sojourns by state, we will study the time of sojourns in the states. The approach is identical to that implemented for the study of the frequency of sojourns by state in the previous section. The results of the statistical processing of the simulated data concerning the sojourn times are summarized in Figure 7.5 and Table 7.4.

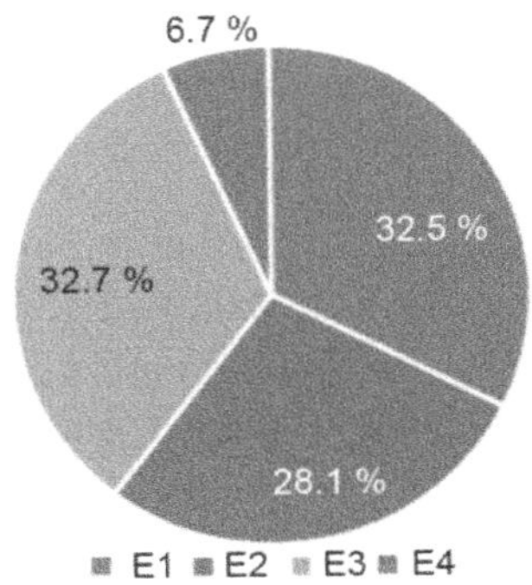


**Figure 7.5.** *Breakdown of sojourn mean times by state. For a color version of this figure, see www.iste.co.uk/elhami/reliability.zip*

| | ***E*1** | ***E*2** | ***E*3** | ***E*4** | ***E*2 + *E*3** | ***E*2 + *E*3 + *E*4** | ***E*1 + *E*2 + *E*3 + *E*4** |
|---|---|---|---|---|---|---|---|
| **($\sigma/\mu$) in %** | 5.1% | 6.9% | 6.5% | 8.8% | 4.7% | 4.6% | 3.5% |
| $\sigma$ | 6.09E+00 | 1.38E+01 | 1.37E+01 | 6.37E+00 | 9.70E+00 | 7.99E+00 | 5.37E+00 |
| $\mu$ **(min)** | 1.18E+02 | 2.00E+02 | 2.11E+02 | 7.21E+01 | 2.06E+02 | 1.75E+02 | 1.51E+02 |
| **No.** | 4.20E+02 | 2.25E+02 | 2.24E+02 | 1.31E+02 | 4.49E+02 | 5.80E+02 | 1.00E+03 |
| **TMS (in min)** | 4.97E+04 | 4.51E+04 | 4.73E+04 | 9.44E+03 | 9.23E+04 | 1.02E+05 | 1.51E+05 |

**Table 7.4.** *Mean sojourn times per $\mu$ state*

We note that the longest sojourn times are observed in the states $E2$ and $E3$ with sojourn mean times close to 200 and 211 min, successively, followed by the state $E1$ with a sojourn mean time of 118 min, and finally the state $E4$ with 72 min. The associated standard deviations are higher than those observed for the frequencies of sojourns, but they remain within the acceptable limits for this application.

We also determined the mean time between transitions corresponding to $10^2 \times 10^3$ transitions which is equal to ~105 min.

It is important to note here that determining the sojourn mean time in each state $E_i$ requires the solution of the following integral:

$$E(T) = \int_0^{\infty} t \cdot dF_i(t), \qquad [7.4]$$

where $F_i(t)$ is the sojourn probability in state $E_i$.

Generally, the sojourn probability function $F_i(t)$ in state $E_i$ is not known. Because of the Monte Carlo method, we have solved equation [7.4] without knowing the probability $F_i(t)$ of being in state $E_i$.

We also find that the sojourn mean time in $E1$ is not the longest compared to other states, even though $E1$ is the most frequent state. If we compare the sojourn mean times in states $E2$, $E3$ and $E4$ to the critical figure of 240 min, which is the failure criterion, we see that the sojourn mean times in states $E2$ and in $E3$ are notably closer to this failure limit compared to the sojourn mean time in the state $E4$. If we add to this the fact that $E4$ is much less frequent than $E2$ and $E3$, it would be logical to conclude that the probability of failure due to a long sojourn in $E2$ or $E3$ is notably greater than the probability of failure due to a long sojourn in $E4$.

To solve this question, we need to determine the probability of failure associated with long sojourns in the states.

### 7.5.6. *The failure probability*

We use the failure criteria already pre-established above: "the process of crisis management fails if the sojourn time in the states $E2$ and $E3$ or in the state $E4$ exceeds 240 min". The simulation campaign previously described provides us with enough information to allow estimating a wide spectrum of quantities and indicators.

To estimate the probability of failure, we already have the histogram of all the transitions between the states (by batches of $10^3$ transitions), the analysis of the frequencies of the transitions by state and the analysis of the sojourn times by state.

The failure probability that characterizes this crisis management process (Table 7.5) is of the order of $1.41 \times 10^{-1}$. The most important contributions are those of the states $E2$ and $E3$ of $\sim 6.8 \times 10^{-2}$ each as previously expected. The contribution of critical sojourns in $E4$ is of the order of $4.6 \times 10^{-3}$. The relative contributions by state are also shown in Figure 7.6.

| Cause of failure | $N \times L$ $(1 \times 1{,}000)$ | $N \times L$ $(100 \times 1{,}000)$ |
|---|---|---|
| **Long sojourn in E2 ($\sigma$/prob %)** | $6.70E-2$ (11.8%) | $6.85E-2$ (0.9%) |
| **Long sojourn in E3 ($\sigma$/prob %)** | $7.80E-2$ (10.9%) | $6.79E-2$ (1.0%) |
| **Long sojourn in E4 ($\sigma$/prob %)** | $9.00E-3$ (33%) | $4.62E-3$ (4.1%) |
| **Long sojourn in E2+E3 ($\sigma$/prob %)** | $1.45E-1$ (7.7%) | $1.36E-1$ (0.6%) |
| **Long sojourn in E2+E3+E4 ($\sigma$/prob %)** | $1.54E-1$ (7.4%) | $1.41E-1$ (0.6%) |

**Table 7.5.** *Estimation of the probability of failure*

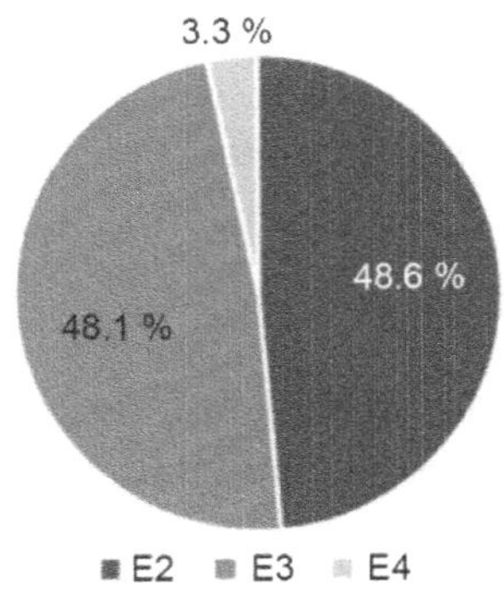


**Figure 7.6.** *The partial contributions of each state (E2, E3, E4) in the failure of the crisis management process. For a color version of this figure, see www.iste.co.uk/elhami/reliability.zip*

It is important to note here that the probability of failure is determined by the following integral:

$$P(t \geq 240) = \sum_{i=1}^{4} \int_{240}^{\infty} dF_i(t), \quad [7.5]$$

where $P(t \geq 240)$ and $F_i(t)$ are the probability of failure and the probability of sojourn in the state $E_i$, respectively. Once again and because of the Monte Carlo method, the integral [7.5] has been determined numerically.

It might also be interesting to determine the share of critical sojourn by state. We have noted in Table 7.6 the number of sojourns by state for a sample size of $10^3$ transitions that occupies in average $\sim 10^5$ min of system operational time. We compare them in Table 7.5 with the numbers of critical sojourns. A critical sojourn corresponds to a sojourn whose duration is equal to or greater than 240 min. We find that only 3.3% of sjourns $E4$ are critical. However, the ratio reaches 30% of critical sojourns for $E2$ and $E3$. The notable low contribution of the state $E4$ to failure comes from the two facts already observed in the previous section: its low frequency of sojourns and the short sojourn mean time (<240 min).

| | E2 | E3 | E4 |
|---|---|---|---|
| **Number of sojourns (/ $10^3$ transitions)** | 232 | 227 | 122 |
| **Number of critical sojourns (/ $10^3$ transitions)** | 69 | 68 | 4 |
| **Ratio (%)** | 30 | 30 | 3.3 |

**Table 7.6.** *Critical transitions and total number of transitions per state*

## 7.6. Simulation of the operation of a redundant system

Systems with 2/3 redundancy have been represented in Chapter 4 by their state graphs and Boolean expression for failure. Therefore, we are not going to simulate the transitions between states because this approach has already been the subject of the previous sections. In this section, we are going to develop an approach that exploits the Boolean expressions of failure or correct operation.

### 7.6.1. *System description*

This is the smoke detection device, already covered in Chapter 4. The device is made up of three independent detectors $(L1, L2, L3)$. The three detectors are based on three different technologies and manufactured by three different manufacturers to eliminate the problems associated with common mode failures. The three detectors are installed in the same workshop hall but at three different positions. Each detector is characterized by an "average time of detection (TMD) – $\tau$".

As in the case dealt with in section 7.5, we start by generating the operation histogram. The basic data needed to generate the operating histogram are the failure rates ($\lambda$), the repair rates ($\mu$) and the start refusal probabilities ( $\gamma$) (Table 7.7).

| # | Component | $\lambda(\mathrm{h}^{-1})$ | $\mu(\mathrm{h}^{-1})$ | $\gamma$ |
|---|---|---|---|---|
| 1 | Detector 1 $(L1)$ | $8.0^{-3}$ | $1.0^{-1}$ | 0 |
| 2 | Detector 2 $(L2)$ | $4.0^{-3}$ | $8.0^{-1}$ | 0 |
| 3 | Detector 3 $(L3)$ | $2.0^{-3}$ | $4.0^{-1}$ | 0 |

**Table 7.7.** *The characteristic parameters of the components*

### 7.6.2. *Operation histogram generation algorithm*

Numerous algorithms can be designed to generate the operating histogram of a system. We present a histogram that seems to us among the easiest, especially because of its clarity.

The following algorithm corresponds to a simulation without acceleration. It generates the successive periods of availability and unavailability of the system. This makes it easy to determine the reliability, the maintainability, the downtime, the mean time to fail and the mean time to repair.

1) Read, N, L, ( $\lambda_j, \mu_j, \gamma_j, j = 1,2,3$).

2) $n = 1$.

3) Initialize the variables:

– component unavailability indicator $I(j) = 0, \quad j = 1,2,3$;

– system unavailability indicator IS(l) = 0 $\forall l$ ;

– the intervals and the time increment $Time(l) = 0, \forall l, \Delta = 0$ ;

– counter $l = 0$.

4) Generate a triplet of variables uniformly distributed in the interval ]0,1[:

$\{\xi_1, \xi_2, \xi_3\}$

5) Update the transition rates:

$\tau_1 = \lambda_1(1 - I(1)) + \mu_1 I(1)$

$\tau_2 = \lambda_2(1 - I(2)) + \mu_2 I(2)$

$\tau_3 = \lambda_3(1 - I(3)) + \mu_3 I(3)$

6) Realize the transitions of the states of the components:

$t_1 = \frac{1}{\tau_1} ln(\frac{1}{1-\xi_1})$

$t_2 = \frac{1}{\tau_2} ln(\frac{1}{1-\xi_2})$

$t_3 = \frac{1}{\tau_3} ln(\frac{1}{1-\xi_3})$

7) Determine the time increment:

$\Delta = \min(t_1, t_2, t_3)$

8) Determine the component whose transition time is minimal:

$j \equiv \min^{-1}(\Delta)$

9) $l = l + 1$.

10) $IS(l) = I(1)I(2) + (1 - I(1))I(2)I(3) + (1 - I(2))I(1)I(3)$.

11) $Time(l) = \Delta$.

12) If $l \geq L$. Continue.

Otherwise:

$if\ j = 1 \quad \rightarrow \quad \mathrm{I}(1) = 1 - I(1)$

$if\ j = 2 \quad \rightarrow \quad \mathrm{I}(2) = 1 - I(2)$

$if\ j = 3 \quad \rightarrow \quad \mathrm{I}(3) = 1 - I(3)$

GoTo 4

13) Statistical processing of the scored $Time(.)$ and unavailability indicators $IS(.)$.

14) Storage of $n^{th}$ sample of a size $L$.

15) If $n \geq N$. Continue.

Otherwise $n = n + 1$.

GoTo 3

16) Statistical processing of $N$samples of a size $L$.

17) STOP.

18) END.

**Algorithm 7.3.** *Operation histogram generation algorithm*

### 7.6.3. *Operation histogram*

We have just generated samples of a size of $L = 10^3$ transitions between the availability and unavailability states of the system. A transition histogram $L = 10^3$ represents an observation mean time equal to ~$38 \times 10^3$ h that varies between $31 \times 10^3$ h and $42 \times 10^3$ h.

A segment of the histogram with emphasis on the time interval [862 h–872 h] is shown in Figure 7.7. The first two system failures occur in this time interval. The first failure occupies the interval [863.4 h–863.6 h] and the second the interval [865.3 h–870.8 h]. There are 36 down intervals out of the $10^3$ simulated intervals.

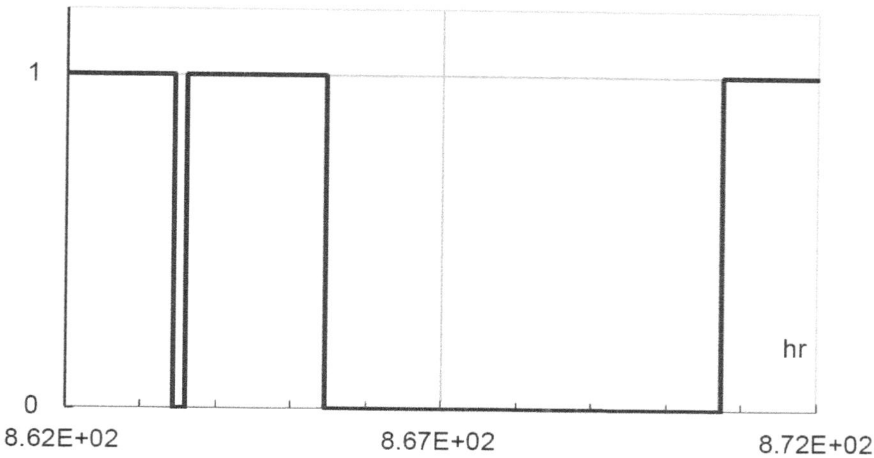


**Figure 7.7.** *A segment of a histogram with focus on the time interval [862 h,872 h]. For a color version of this figure, see www.iste.co.uk/elhami/reliability.zip*

### 7.6.4. *The mean time to fail*

A campaign of simulations of $10^2$ batches of $10^3$ transitions each, allowed us to obtain sufficient data with a good statistical quality regarding the estimators of the time to fail. Correct functioning is defined by the variables Time(.) whose indicators $IS(.)$ have a zero value. The appropriate statistical treatment of these variables demonstrates that the set of these random variables is represented by the distribution illustrated in Figure 7.8. The distribution is characterized by a mean time to fail (MTTF) equal to $\sim 2.0 \times 10^3$h, a most probable time to fail equal to $1.73 \times 10^3$ h, a variance of 14.8 h, a lower bound of $1.05 \times 10^3$ h and an upper bound equal to $4.95 \times 10^3$ h.

If, and only if, the failure rate ($\lambda_{\text{sys}}$) of the redundant system was constant over time, we could have written:

$$\lambda_{\text{sys}} = (\text{MTTF})^{-1} = (1.91 \times 10^3)^{-1}\ \text{h}^{-1}.$$

We already know from the example treated in Chapter 4 that the failure rate ($\lambda_{sys}$) of the system is not constant over time. Because we know the exact analytical expression that determines this failure rate with the time because of the model developed in Chapter 4. This analytical expression is indeed complex. However, we could determine the time integral of this failure rate expression using the trapezoid technique in Chapter 4.

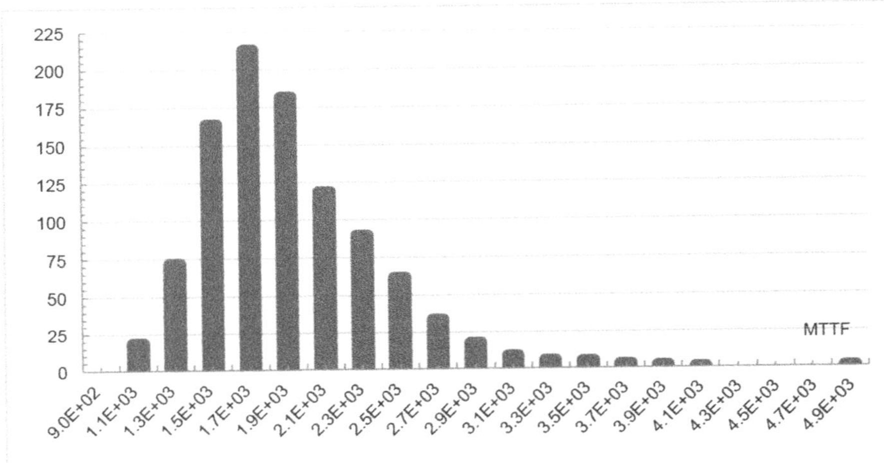


**Figure 7.8.** *Failure occurrence frequency vs. time-to-fail estimators [$10^2 \times 10^3$]. For a color version of this figure, see www.iste.co.uk/elhami/reliability.zip*

Determining the integration over time of the failure rate is necessary for the determination of reliability or the reliability of the system, whether the failure rate is constant or time dependent. The use of Monte Carlo simulation for modeling dynamic systems is generally a complex subject. However, our case is a simple one.

Remember that the reliability $F(T)$ of any system is determined by:

$$F(T) = \text{Exp}(-\int_0^T \lambda_{sys}(\xi)d\xi).$$

Let us also recall that knowing the expression for the reliability of the system is necessary to determine the MTTF, which is nothing other than the mathematical expectation of the uptime:

$$(\text{MTTF}) = \int_{\xi=0}^{\infty} \xi dF(\xi).$$

In this section, the simulation by the Monte Carlo method has allowed us to determine numerically the MTTF without any a priori knowledge of the analytical expression of system reliability $F(T)$.

But how can we finally determine the reliability by Monte Carlo simulation technique?

### 7.6.5. *Algorithm for determining system reliability*

We must determine the reliability of the system without any a priori knowledge neither of the density probability function of failure occurrence nor of the analytical expression of the failure rate of the system. This will be achieved by the statistical exploitation of the data already generated by the simulation of the above operation histogram.

We recall that the proper operation is defined by the variables $Time(.)$, whose indicators $IS(.)$ have a zero value. To note that we have $N$ samples of size $L = 10^3$ simulations each.

The general algorithm is composed of seven phases, after having identified the maximum value of $Time(.)$ in the set of $N \times L$ data produced by the simulation.

In this algorithm, we determine the unreliability instead of the reliability. The reason comes essentially from the fact that the quantity that represents the unreliability is smaller for small intervals of time and increasing over time, while the reliability evolves over time in the opposite way.

Generally, it is better to score a quantity that increases over the observation time than a quantity which decreases. The statistical quality of the scored data increases with the accumulation of time.

The algorithm is described as follows:

1) Divide the interval $[0, Time_{\max}(.)]$ into $K$ subintervals of the same length and determine the center of each interval:

$$t(k);\ \forall k \in [1, K]$$

2) Count the number of failures $C(k,n)$ within each subinterval $k$ $(k; \forall k \in [1, K])$, using the data $Time(l)$, $(l;\ \forall l \in [1, 10^3])$, of the sample $n$.

3) Determine the scored probability of failure $W(k,n)$ in each subinterval:

$$W(k,n) = C(k,n) \times 10^{-3}$$

4) Return to step 2 as many times $n$ as there are samples:

$$n = 1,2, \dots, N$$

5) Determine the mean value of the probability of failure, also called the estimator $\overline{W}(k)$, for subinterval $k$:

$$\overline{W}(k) = E\big(W(k,n)\big) = \frac{1}{N}\sum_{n=1}^{N} W(k,n)$$

6) Trace the failure probability estimator $\overline{W}(k)$ within $t(k);\ \forall k \in [1, K]$ to obtain the reliability of the system over time.

7) Trace reliability $F(k)$ over time $t(k)$, where:

$$F(k) = 1 - \overline{W}(k)\ ;\ \forall k \in [1, K]$$

**Algorithm 7.4.** *Monte Carlo simulation of the unreliability of a system 2/3*

### 7.6.6. *System reliability estimation*

The application of the algorithm described above allowed the determination of the reliability profile of the system over time by statistically processing the variables $Time(.)$. We have $N$ samples of one size $L$ simulations each, where $N = 10^2$ and $L = 10^3$.

As a reminder, the unreliability $D(t)$ plotted in blue (Figure 7.9) is the numerical solution of this expression:

$$D(t) = 1 - \mathrm{Exp}\left(-\int_{\xi=0}^{t} \lambda_{\mathrm{sys}}(\xi)d\xi\right).$$

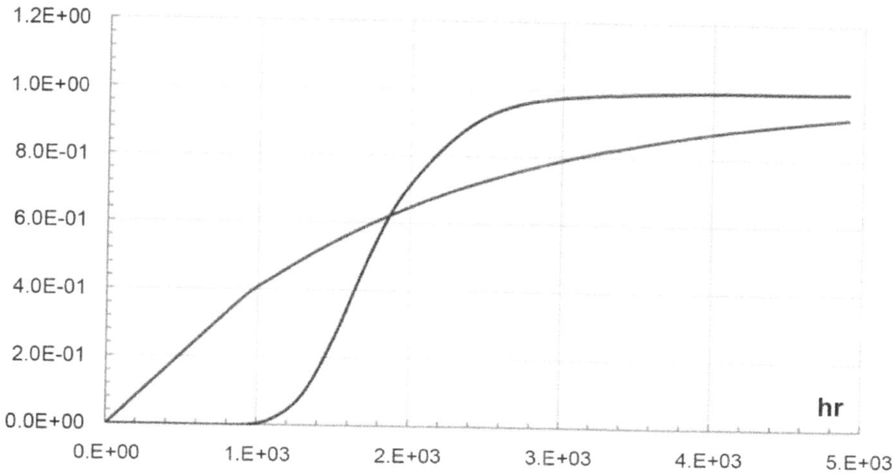


**Figure 7.9.** *Comparison between the profile of unreliability estimated by the simulation (blue) and the unreliability if $\lambda_{sys} = (1.91 \times 10^3 h^{-1})^{-1}$(red). For a color version of this figure, see www.iste.co.uk/elhami/reliability.zip*

The complexity of the expression describing the failure rate $\lambda_{\text{sys}}$ was reported in Chapter 4. Failing to find an analytical solution to the integral in the above equation, we must determine the unreliability numerically.

The profile of the variation in the unreliability over time is shown in blue (Figure 7.9). It is compared with the $\lambda_{\text{sys}}$ unreliability profile, calculated if the constant system failure rate is equal to $(1.91 \times 10^3)^{-1}\ \text{h}^{-1}$. We see a noticeable difference.

This noticeable discrepancy between the results of Chapter 4 and this chapter sheds light on the difficulty of simulation dynamic systems by direct unbiased Monte Carlo simulation. Dynamic systems simulation requires the development of more complex simulation algorithms designed on a case-by-case basis exploiting biasing and acceleration techniques. This subject is outside the didactic framework of the book.

## 7.7. Conclusion

The Monte Carlo method is a sampling method that is universally used to solve complex multidimensional integrals numerically. It is used in several fields of application ranging from quantum mechanics and particle physics to structural mechanics and system reliability.

Its use in system reliability and operational safety allows us to deal with complex systems either by the number of their components, by the multitude of their operational phases or by the change in the structure of the system according to its operating phase. This also makes it possible to take regulatory and organizational constraints into consideration.

The Monte Carlo method is also an efficient tool for calculations and propagation of uncertainties.

The method finds its full power when dealing with stationary systems described by multidimensional integrals.

# 8

# Physical Tests used in Reliability

## 8.1. Introduction

Long since, reliability tests were conducted to test the proper functioning of manufactured objects before commercial distribution. Over time, testing methods and techniques were created and then standardized. This development has allowed a considerable increase in the reliability of the manufactured objects.

Accelerated tests have been developed to estimate the lifespan of a product without waiting as long time as the product actual operational lifetime. The increase in the reliability of industrial products has led to an increase in lifespan such that it has become increasingly difficult to detect defective products within a reasonable time interval, using simple testing functions or conventional accelerated tests.

Other tests have then to be set up. These are the aggravated tests. Indeed, the greater the reliability, the later the malfunctions appear in the life of the object. It is then necessary to reveal the defects of the object to make it more reliable and to go even beyond the future conditions of use.

These tests make it possible to obtain data on the reliability of the designed or manufactured systems. When a new prototype includes innovations and there are no failure data yet or they are unavailable, it is necessary to build a knowledge base using physical tests to optimize the system's design and/or operational performances. This makes it possible to estimate the reliability of the systems as a function of the stresses which they will be subjected to.

Among the different tests used to test the reliability of industrial products, we can distinguish between accelerated tests, aggravated tests and Bayesian-based tests. Accelerated and aggravated tests have one common factor. That is their short testing duration compared to the lifespan of the tested products. But they have two different purposes. Bayesian-based tests make it possible to increase the relevance of the failure probability through coupling the use of the Bayesian theory and the outcome of few tests. The reliability improvement cycle is presented in Figure 8.1.

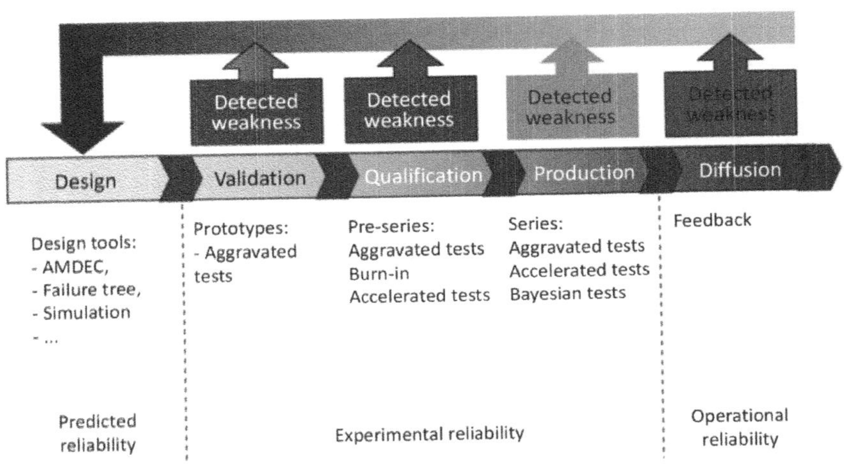


**Figure 8.1.** *Reliability improvement cycle. For a color version of this figure, see www.iste.co.uk/elhami/reliability.zip*

Reliability testing generally takes place at the beginning of a product's life cycle and is divided into three stages [DAM 21]:

– *Predicted reliability estimate*: from the beginning, and throughout the design, tools are used (APR, FMECA, fault trees, reliability diagram, etc.) to estimate the reliability of the future object. Predicting reliability includes all engineering activities to *predict* the reliability performance of the final system. There is not yet an integrated solution to virtually simulate and estimate the reliability intervals of a future system.

– *Experimental estimate*: as soon as the first prototypes are manufactured, it is possible to carry out tests. These will show design and manufacturing flaws, as well as product specification margins (Bayesian testing, highly accelerated life test – HALT). Once the necessary corrections have been made, a pre-series can be subjected to a burn-out test, eliminating the defective elements (HASS, highly accelerated stress audit – HASA). Experimental reliability testing uses all the tools and test methods to *demonstrate* the reliability of the system.

– *Operational estimate*: once the product has been distributed, an estimate of its operational reliability is made based on operational feedback data. Operational reliability *verifies the* reliability performance of the system in its real life by analyzing real field data.

We can present the phases of the V-cycle by associating the phases of the reliability development [LED 19] as shown in Figure 8.2.

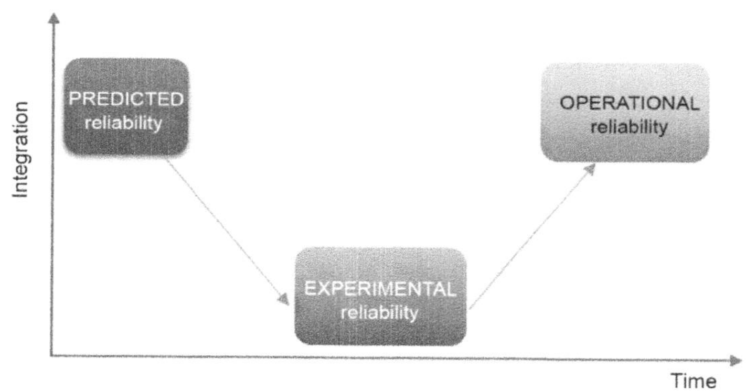


**Figure 8.2.** *The reliability cycle. For a color version of this figure, see www.iste.co.uk/elhami/reliability.zip*

## 8.2. Accelerated tests

Accelerated testing makes it possible to prematurely age a system in operational conditions similar to the normal ones. Several models, mostly empirical, exist. Accelerated trials include:

– HAST (highly accelerated stress test): this test can reduce the test duration from 1,000 to approximately 100 h through biasing the test temperature and humidity (THB – temperature humidity bias);

– STRIFE (stressful life test): this is a test reproducing the maximum stress operational conditions.

### 8.2.1. *The different acceleration laws*

These different empirical laws are used to predict the future behavior of the systems and are based on experimental observations. They each have a restricted field of application and must be used with discernment. They have been transcribed from previous studies (ELH 13b).

#### 8.2.1.1. *Arrhenius model*

It is used to model the aging of a component subjected to oxidation or diffusion phenomena at given temperature. With this law of acceleration, we can simulate 20 years of aging in 1,000 h (about 42 days):

$$V = A.e^{\left(-\frac{B}{T}\right)}, \tag{8.1}$$

with $A$ being a constant and $B = \frac{e \times v_a}{K}$, $V$ is the reaction rate, $T$ is the temperature, $e$ is the coulomb constant, $v_a$ is the activation potential and $K$ is the Boltzmann constant.

We define an acceleration factor $FA$ as follows:

$$FA = \frac{\tau_1}{\tau_2} = e^{B\left(\frac{1}{T_1}-\frac{1}{T_2}\right)}, \tag{8.2}$$

with $\tau_i$ lifetimes and $T_i$ temperatures.

The disadvantage of this model is that it does not take into account other phenomena, such as humidity, which can influence the aging of the system studied.

EXAMPLE 8.1.–

The activation energy of oxidization for the steel is 0.55 eV. Let us calculate the acceleration factor of steel subjected to a temperature of 100°C during a test, instead of 30°C under normal conditions.

$$FA = e^{\left[\frac{e \times v_a}{K}\left(\frac{1}{T_1}-\frac{1}{T_2}\right)\right]}, \tag{8.3}$$

$$FA = e^{\left[\frac{1.6.10^{-19}\times 0.55}{1.38.10^{-23}}\cdot\left(\frac{1}{303}-\frac{1}{373}\right)\right]}. \qquad [8.4]$$

$FA \approx 36.2$, that is, aging 36 times faster than in normal conditions.

### 8.2.1.2. *Peck's model*

It is an "extension" of Arrhenius law by adding a term to take humidity into account.

This model is used when test temperature and humidity must be taken into account. It is particularly suitable for electronic components and conductors:

$$\tau = A\left(\frac{H_0}{H_a}\right)^{-m} e^{\left(\frac{E_a}{kT}\right)}, \qquad [8.5]$$

with:

- $T$: absolute temperature in $K$;
- $H$: percentage of humidity;
- $k$: Boltzmann constant;
- $E_a$: activation energy in eV;
- $m$: model parameter.

### 8.2.1.3. *Inverse power or Coffin–Manson model*

An inverse power model is used for specific stresses when the stress level is constant. If the stress is the temperature, this law becomes a particular case called "Coffin–Manson law". This law is used for bearings, lamp filaments and so forth

$$\tau = \left(\frac{A}{V}\right)^{\gamma} \qquad [8.6]$$

with:

- $\tau$: expected service life;
- $V$: constant stress.

$A$ and $\gamma$ are the failure stress and the test constants, respectively. The acceleration factor $FA$ is then given by:

$$FA = \frac{\tau_1}{\tau_2} = \left(\frac{V_2}{V_1}\right)^{\gamma}. \qquad [8.7]$$

Example for a bearing:

$$L_{10} = \left(\frac{C}{P}\right)^{n}. \qquad [8.8]$$

This formula expresses the service life $\tau$ of a bearing in millions of revolutions $n$, according to a reference load $p$, and $C$ is the test constant (dynamic load capacity).

EXAMPLE 8.2.–

Documentation from a manufacturer gives us $C$ a dynamic load capacity for a radial contact bearing of 5,200 daN. In reality, the equivalent load actually applied is 1,400 daN. The coefficient $n$ is 3 for the type of bearing used:

$$L_{10} = \left(\frac{5{,}200}{1{,}400}\right)^{3} = 51.2\ (\times\ 10^{6}).$$

The bearing will withstand, at this load, 51.2 million revolutions (for a survival rate of 95%).

#### 8.2.1.4. *Eyring model*

The Eyring model is used for damage mechanisms where mechanical stress and temperature stress are combined:

$$\tau = \left(\frac{A}{T}\right) e^{\left(\frac{B}{kT}\right)} . e^{\left(V\left(C+\frac{D}{kT}\right)\right)}, \qquad [8.9]$$

with:

– $T$: absolute temperature in $K$;

– $V$: stress value;

– $k$: $8.6171.10^{-5}$ eV/K Boltzmann constant;

$A$, $B$, $C$ and $D$ are failure and test dependent constants.

Proportional random model:

$$r_\lambda = \frac{\lambda(t,z_A)}{\lambda(t,0)} = g(z). \tag{8.10}$$

In general, $g(z)$ is of exponential form, hence:

$$r_\lambda = e^{(\sum_i b_i \times z_i)} = e^{(b^t.z)}. \tag{8.11}$$

The acceleration factor for Weibull's law then becomes:

$$F_A = e^{\left(\frac{b^t.z}{\beta}\right)}. \tag{8.12}$$

The coefficient $b$ is used to parameterize the model; it is possible to use the partial likelihood method or multiple regressions.

## 8.3. Simple mechanical fatigue, Wöhler model

There are many models associated with the phenomenon of fatigue. We can cite the models of Bastenaire, Lemaître and Chaboche, Stromeyer or Basquin. They express the life stress cycle as a function of the maximum stress σ. On the Wöhler curve (Figure 8.3), there are three domains:

– *Domain D1*: low-cycle plastic fatigue under high stress. Failure occurs after a small number of cycles and significant plastic deformation.

– *Domain D2*: limited fatigue or limited endurance. Fatigue at a high number of cycles.

– *Domain D3*: unlimited endurance, where there is never fatigue failure, regardless of the number of cycles.

The Wöhler curve is the best known of the representations of mechanical fatigue models; it is also called the stress-cycle number (SN) curve.

These curves are plotted from the compilation of the results of numerous tensile or torsion tests on standardized specimens. The specimens are subjected to an alternating stress.

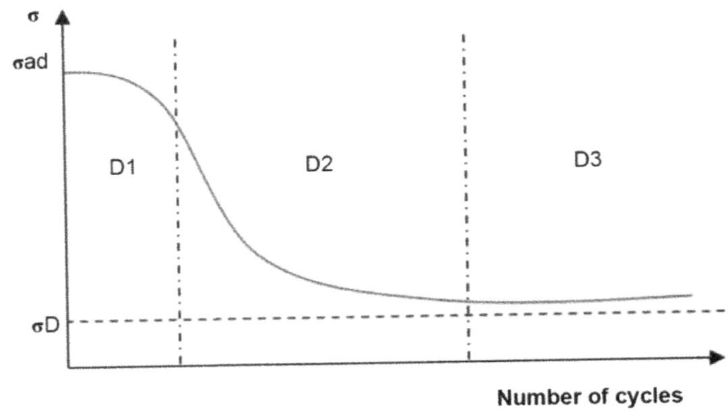


**Figure 8.3.** *Wöhler curve. For a color version of this figure, see www.iste.co.uk/elhami/reliability.zip*

The curve expresses the stress resistance of a material as a function of the number of cycles for a given stress. These curves are generally established for a survival of 50% of the specimens. There are generally three phases:

– *Low-cycle fatigue (domain D1)*: when the specimen is subjected to a small number of cycles, its resistance is slightly affected and is still equal to the nominal resistance of the material.

– *Fatigue with a large number of cycles (domain D2)*: if the specimen is subjected to a large number of cycles, its resistance decreases sharply.

– *Endurance limit (domain D3)*: in this part of the curve, if the stress is never greater than a certain level, the life time is expected to be unlimited. However, this endurance limit does not exist for all materials.

In the case of a fatigue failure, when the failure occurs at stress level below the elastic limit, the mode of failure is due to a crack in the macroscopic structure of the material. This will gradually grow, reducing the resistance and increasing the stress in the material. These cracks generally start at a particular place in the metal in the vicinity of inclusions. An inclusion is a defect in the structure of the material due to a concentration of stress. Figure 8.4 shows the Wöhler curve with separation of crack initiation and crack propagation.

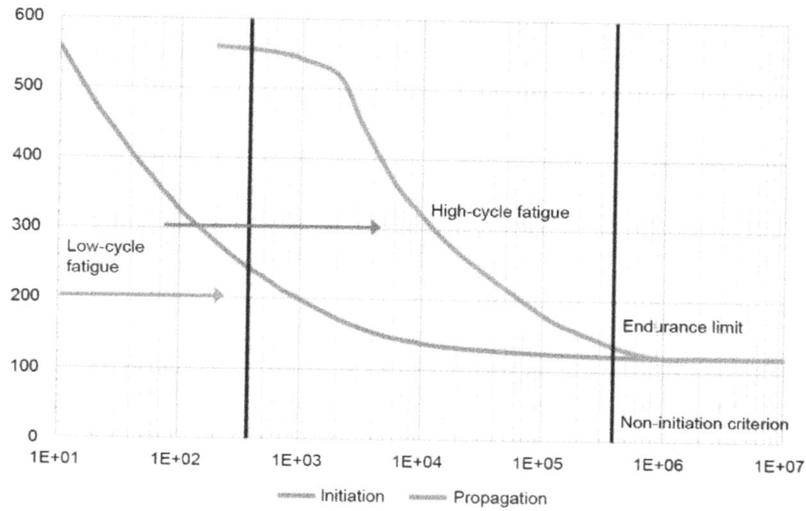


**Figure 8.4.** *Wöhler curve with separation of crack initiation and crack propagation. For a color version of this figure, see www.iste.co.uk/elhami/reliability.zip*

NOTE.– These tests are expensive, and they are often carried out for a small number of materials, and then extrapolated for the same family of materials. It is necessary to carry out a large number of tests because local variations in the material are common and can cause great disparities in the test results. It is possible to couple these tests with a temperature parameter; one then obtains a response surface which depends on the stress, the number of cycles and the temperature. These tests are even rarer and more expensive.

### 8.3.1. *Empirical laws of fatigue*

#### 8.3.1.1. *Basquin's law*

This law is used for cases of high-cycle fatigue (stress well $\sigma_{max}$ below the elastic limit):

$$\Delta\sigma . {N_f}^a = C_1, \qquad [8.13]$$

where $\Delta\sigma$ is the cyclic stress amplitude with an average stress equal to zero, $N_f$ is the number of cycles to failure and $C_1$ and $a$ are material – dependent constants.

#### 8.3.1.2. *Coffin–Manson law*

This law is used for low-cycle fatigue (number of cycles less than 50,000 and solicitation stress close to the elastic limit):

$$\Delta\varepsilon_{plast} \cdot {N_f}^b = C_2, \quad [8.14]$$

where $\Delta\varepsilon_{plast}$ is the plastic strain amplitude, $N_f$ is the number of cycles to failure, $b$ and $C_2$ are material – dependent constants.

#### 8.3.1.3. *Miner's rule*

It is a rule that makes it possible to estimate the fatigue of a material by accumulation of the damage.

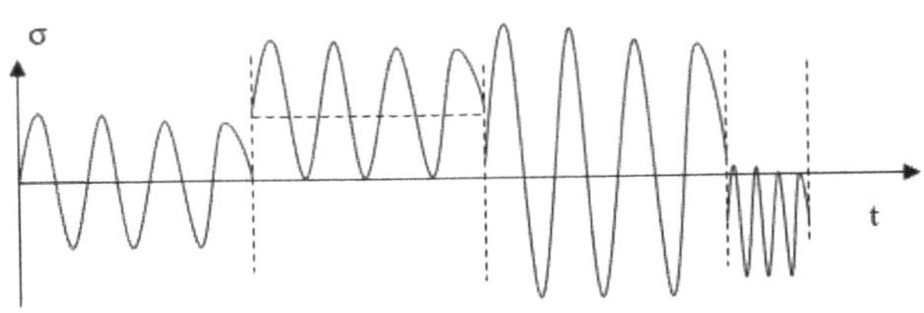


**Figure 8.5.** *Example of solicitation cycle*

This amounts to calculate the percentage of damage created by each stress cycle and calculating the sum. This method does not take into account the order of the stress cycles, which can, in certain cases, give completely false results.

$$\sum_i \frac{n_i}{Nf_i} = 1, \quad [8.15]$$

where $n_i$ is the number of cycles performed and $Nf_i$ is the number of cycles required for failure.

## 8.4. Aggravated tests

Aggravated tests are performed on objects tested at stresses greater than their real operational ones, that is, they go beyond the manufacturer's specifications.

In addition to being accelerated, these tests are used to test the limits of a material. They make it possible to reveal structural manufacturing and design defects. The problems encountered will be analyzed and corrected, thus increasing the reliability of the tested systems.

Various techniques have been used for a long time (about 30 years) in the United States but are relatively recent in France (about 15 years) and struggling to get established.

In order to be more effective, they must be integrated into the product development process. Properly used, they make it possible to minimize and correct the major defects in a system, in order to optimize production techniques, limit returns, and increase the operational condition limits to approach as close to the technological limit as possible. Initially, these methods were exclusively used in electronics before spreading to automotive and aviation industrial sectors. They are applicable to any mechatronic component as well. Mechanical components can potentially benefit from these methods. However, the pre-existing mechanical culture in favor of destructive tests limits the dissemination of these tests.

There are large number of tests, all have in common a strong emphasis on loading modes and loading sequences.

### 8.4.1. *The main aggravated tests*

#### 8.4.1.1. *Highly accelerated life test*

It was Gregg Hobbs who, in 1988, first proposed a family of aggravated tests called HALT for Highly Accelerated Life Test [YAI 18]. The meaning of the acronym was changed in 2013 during IEC 62506 (2013) by highly accelerated limit test. In the development cycle of a system, the strategy of HALT aggravated tests makes it possible to quickly reach design maturity with a high level of reliability (Figure 8.6).

The failure of a more or less complex system is always due to the application of a threshold constraint (mechanical, electrical, thermal, vibrational, climatic, etc.), which exceeds the intrinsic resistance of the system [ELH 13b].

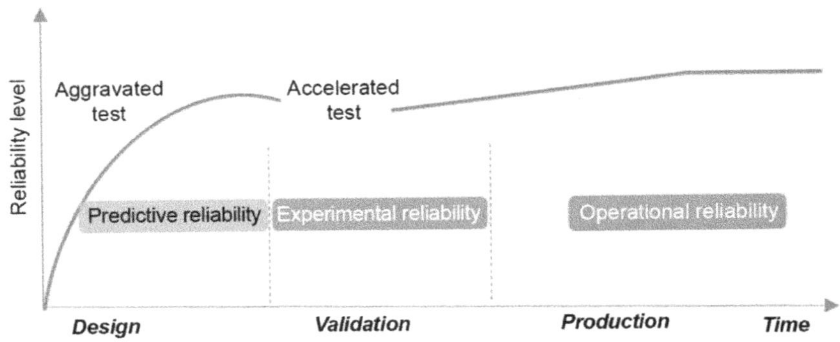


**Figure 8.6.** *Reliability growth by aggravated tests. For a color version of this figure, see www.iste.co.uk/elhami/reliability.zip*

The purpose of HALT tests is to determine the operating limits of each parameter and coupled parameters. One or more stresses are applied in stages, while the apparatus is in operation. On the one hand, we check that the device complies with the standards. On the other hand, we obtain the contours separating between the safe and the unsafe functioning.

Safe functioning is the area of operation where the manufacturer's recommendations are complied with, and robustness is the margin space between the manufacturer's specifications and the degraded function. The degraded function is the area where the product still works, but badly, because if we return to so-called normal operating conditions, the product will work correctly again. Beyond the degraded function is destruction. No return-to-normal is possible.

It is possible to vary the temperature (negatively, as well as positively) to stress using accelerations or vibrations and to combine different stresses (tension, temperature, electromagnetic waves, etc.).

The stresses that are usually the most stressful for the material are, in ascending order, thermal cycling, random vibrations and finally the vibration-thermal coupled cycling. It is important to adapt the test to the system and to the type of defects that one wishes to detect because materials can be sensitive to certain stresses. The stress types should respect the predicted failure mode scenarios and be representative of the types of future operating conditions.

It is not possible to test a system at random.

If a failure mode is not considered, the test output becomes invalid and the designers may not detect an error or a defect which may only be discovered by the user.

#### 8.4.1.2. *Burn-in tests*

HASS (highly accelerated stress screening) is an accelerated and aggravated burn-in test. To be effective, stress thresholds must be determined using a HALT test. Burn-in is a stimulation operation allowing early failures to be detected, without altering the lifespan of healthy components. HASS is not just a test, it is a process.

The HASA is a test of burn-in by statistical sampling. It requires a perfect mastery of the manufacturing process on a stabilized production. It requires the information obtained with the HALT and HASS tests.

ESS (environmental stress screening) is mainly used in electronics. It is a burn-in test; it allows equipment to be tested and "premature mortality" to be reduced. The materials are repaired or removed from the production line.

The implementation of these tests is very expensive and leads, most of the time, to the destruction of the products. All these tests allow the detection of defects, but it is necessary to find the origin with high certainty in order to make good corrections. The proof of screen (POS) is the test of burn-in. This is a step that demonstrates that burn-in does not damage good quality equipment and reveals the defects of other devices tested.

Due to the cost of these tests, these tests are carried out mainly on special equipment, where reliability issue is critical, such as on-board equipment or on large series, which will make it possible to amortize the cost of the study and the modifications, and to significantly reduce warranty returns and user's technical support.

#### 8.4.1.3. *Example of test cycles in aggravated tests*

The first cycles are carried out to delimit the contours of the safe functioning area of the product according to the stresses and combinations of stresses and the suspected failure modes. A burn-in test can then be performed.

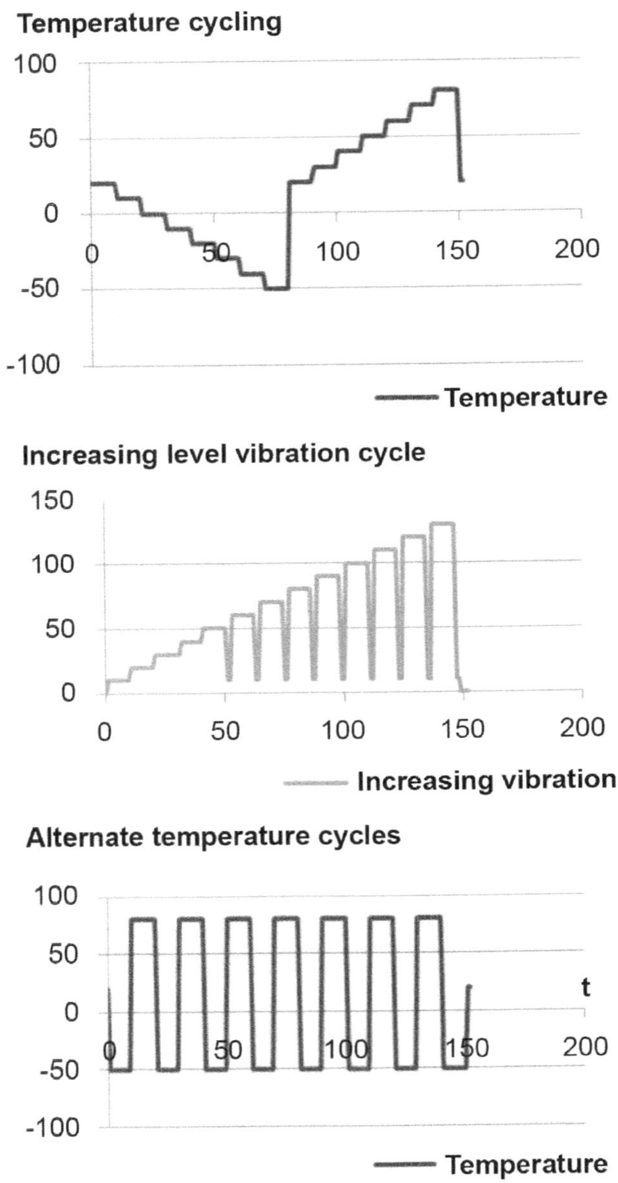


**Figure 8.7A.** *Different types of aggravated test cycles. For a color version of this figure, see www.iste.co.uk/elhami/reliability.zip*

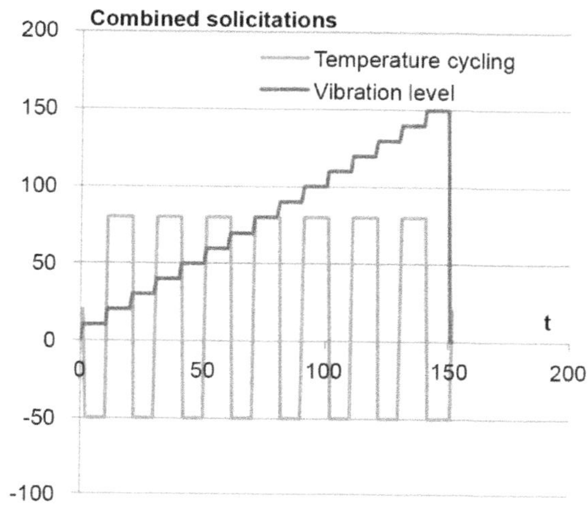


**Figure 8.7B.** *Different types of aggravated test cycles (continued). For a color version of this figure, see www.iste.co.uk/elhami/reliability.zip*

Figure 8.8 shows a test chamber, allowing HALT and HASS tests to be carried on.

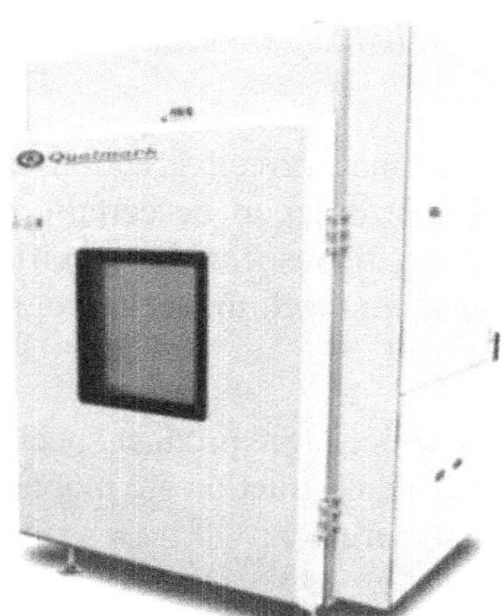

**Figure 8.8.** *Aggravated test enclosures (Qualmark). Thermal environment: −100°C to +200°C, temperature variation of 70°C/min, cryogenics by liquid nitrogen, random vibrations six degrees of freedom between 10 and 10,000 Hz up to 100 g. For a color version of this figure, see www.iste.co.uk/elhami/reliability.zip*

To be able to process data and obtain sufficiently reliable extrapolations, it is necessary to know the model of reliability law that applies to the studied system.

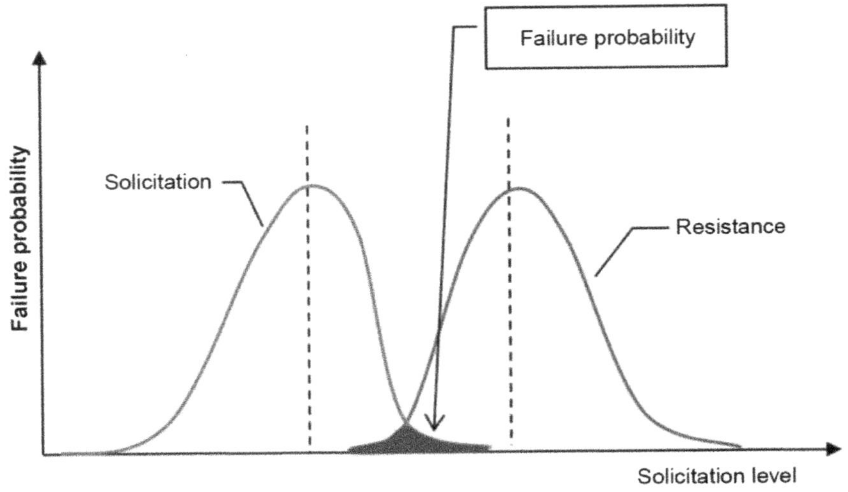


**Figure 8.9.** *Determination of failure probabilities with the statistical distributions of the stresses and the resistance of the system. For a color version of this figure, see www.iste.co.uk/elhami/reliability.zip*

The interest of the tests and corrective actions is to obtain the widest possible operating margins in order to be certain, despite the vagaries of manufacture and distribution, to respect the specifications of the product. The operating margins must approach the technological limit, the ideal case being when both are combined.

When the product is optimized, the product specification is reduced. The margins between operation and destruction are more or less large and depend above all on the operational margin.

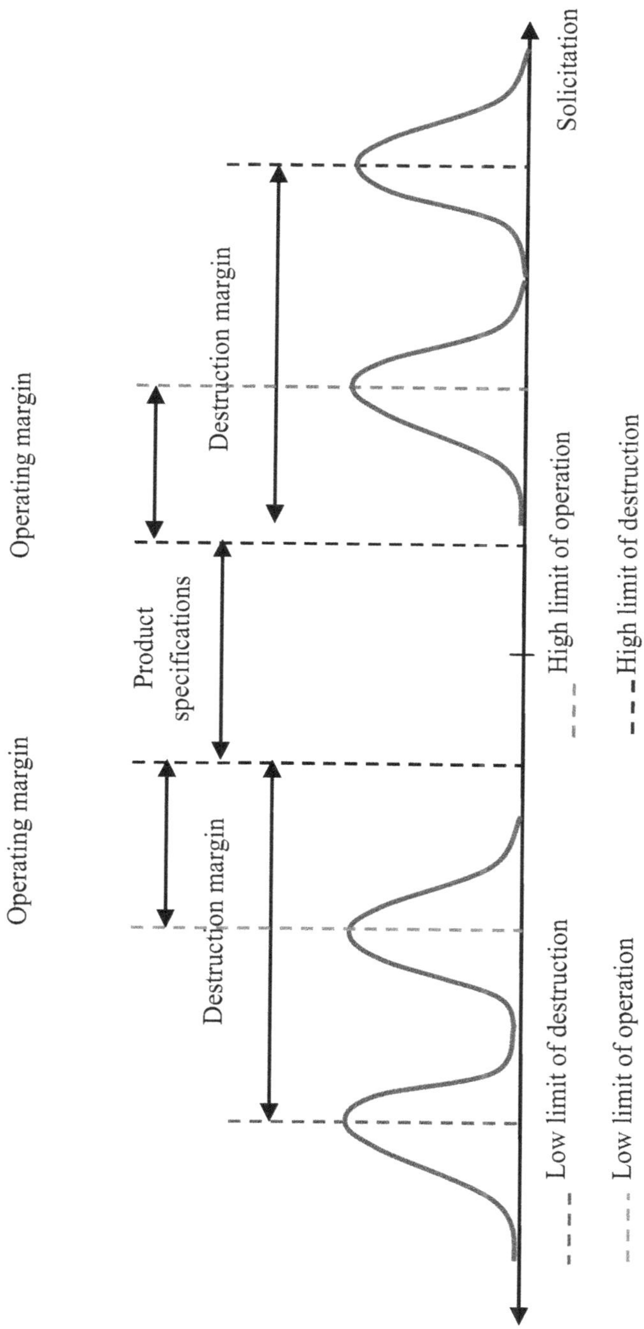


**Figure 8.10.** *Level of operating margins before optimization. For a color version of this figure, see www.iste.co.uk/elhami/reliability.zip*

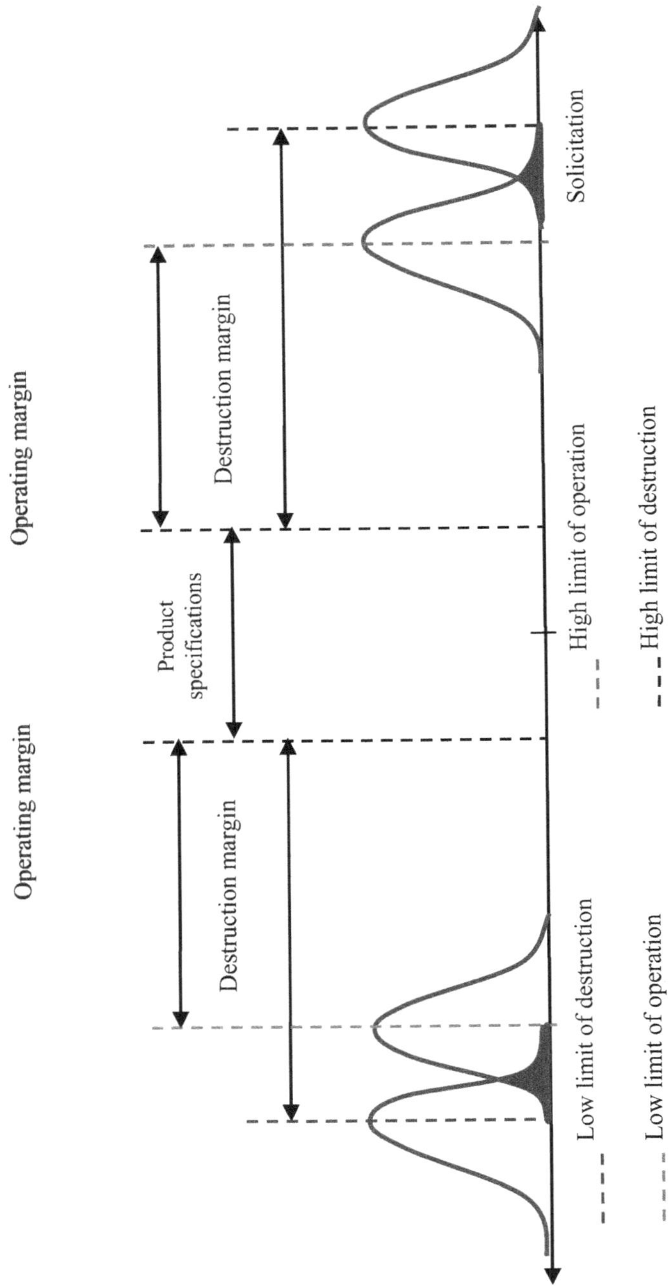


**Figure 8.11.** *Margins after modifications following an aggravated test. For a color version of this figure, see www.iste.co.uk/elhami/reliability.zip*

When the product is tested and improved, the product specifications are still the same, but the margins between operation and destruction become greater. They are also closer to each other. That is to say that a significant part of the production will be destroyed before undergoing malfunctions. What seems to be a disadvantage is not one. First, this limit is beyond the old operating margin; then, the majority of the products would not have operated at these stress levels anyway, and finally a significant part will still operate when it would have malfunctioned without these modifications.

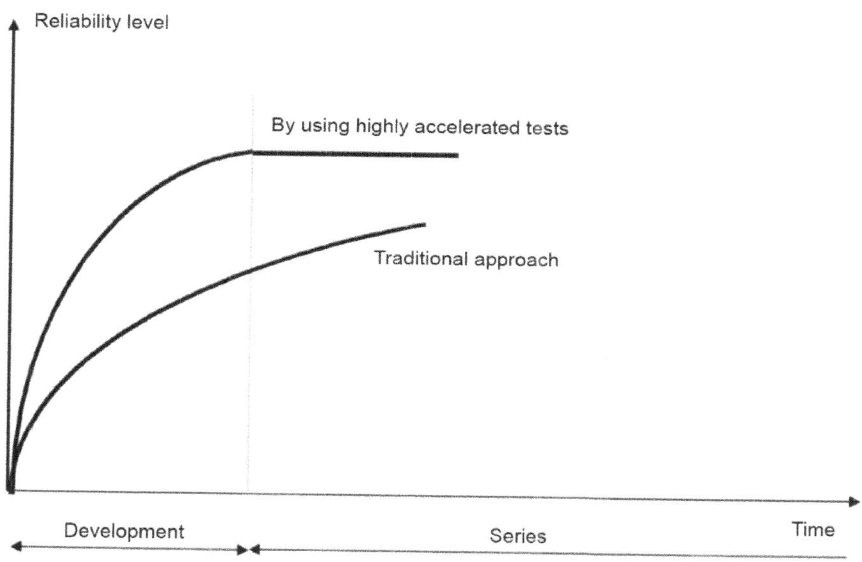


**Figure 8.12.** *Evolution of reliability in development and production*

NOTE.– The intelligent practice of aggravated tests must allow us to:

– accelerate the maturation of a product in development and therefore be able to deliver mature products from the start of the series;

– improve the reliability and durability of the product in the field, via its increased robustness and the elimination of manufacturing defects;

– reduce development times and costs;

– make burn-in more effective in revealing manufacturing defects;

– significantly reduce the number of retakes after qualification;

– better meet customer expectations;

– improve the brand image of the company.

Figure 8.11 shows the acceleration of the reliability performance maturation that can be obtained at the end of development, because of the performance of highly accelerated tests, compared to the traditional approach.

## 8.5. Bayesian tests

These tests require both knowledge of the studied phenomenon (its driven parameters, history, base of experience) and a limited number of real tests. That makes it possible to refine a theoretical statistical model using a reduced number of tests.

This type of testing has three stages.

First, it is necessary to compile as much data as possible on the studied object and to create a theoretical reliability model with all the available data (a priori law). Then, we have to carry out a real test and collect data on a few samples and finally we have to "post-process" the results (a posteriori law) of the tests to refine the theoretical model using Bayes' theorem:

$$g(\theta|t) = \frac{f(T|\theta).g(\theta)}{\int_{D(\theta)} f(T|\theta).g(\theta).d\theta}, \qquad [8.16]$$

with:

– $\theta$: vector of the unknown parameters of the lifetime law (failure rate, repair rate, etc.);

– $T$: test results (times of failure, etc.);

– $f(T|\theta)$: likelihood function of the sample;

– $g(\theta|T)$: a posteriori probability density of the parameter $\theta$ (conditional probability density depending on the available information);

– D($\theta$): domain of definition of the random variable.

It is possible to estimate the parameter $\theta$, as well as the optimal size of the sample to be tested.

Indeed, as soon as the number of statistical tests increases, Bayes' theorem is less and less valid and the applicable law approaches that of large numbers.

## 8.6. Fatigue damage analysis

### 8.6.1. *Formulations and development*

The primary purpose of multiaxial fatigue criteria is to determine the location of the critical point in a structure. Many authors may think that crack initiation is governed by the second invariant of the deviating stress tensor [BEN 15]. This is visible in the Crossland criterion (equation [8.17]) and the Sines criterion (equation [8.18]) where $J_{2,a}$ is the second invariant of the stress deviator:

$$D_{\text{Crossland}} = \frac{\sqrt{J_{2,a}} + \alpha \max\limits_{0 \le t \le T}\{p(t)\}}{\beta} \le 1, \quad [8.17]$$

$$D_{\text{Sines}} = \frac{\sqrt{J_{2,a}} + \alpha\, \mathrm{E}[\mathrm{p(t)}]}{\beta} \le 1, \quad [8.18]$$

where $p(t)$ is the hydrostatic pressure (equation [8.19]), and $\alpha$ and $\beta$ are material constants. For the criterion of Sines, they are defined by the limit of endurance in tension alternating $f_{-1}$, the endurance limit by alternating torsion $t_{-1}$ and breaking strength $Rm$ (equations [8.20] and [8.21]):

$$p(t) = \frac{1}{3}\left(S_{xx} + S_{yy} + S_{zz}\right), \quad [8.19]$$

$$\alpha = \frac{3t_{-1}(R_m + f_{-1})}{f_{-1}R_m} - \sqrt{6}, \quad [8.20]$$

$$\beta = t_{-1}. \quad [8.21]$$

It can be noted that the Crossland criterion requires the calculation of the maximum of the hydrostatic pressure, whereas the Sines criterion uses its average. This reduces the damage calculation time if it is calculated according to the Sines criterion and has almost the same result as the Crossland criterion [YAI 18].

Now, consider a linear structure subjected to stationary ergodic Gaussian loads. The Sines criterion is then defined by:

$$E[D_{\text{Sines}}] = \frac{E\left[\sqrt{J_{2,a}}\right] + \alpha E[p(t)]}{\beta} \leq 1. \qquad [8.22]$$

The problem, now, is in calculating the average shear stress $E\left[\sqrt{J_{2,a}}\right]$. Indeed, for a proportional loading, $\sqrt{J_{2,a}}$ can be defined by equation [8.23]:

$$\sqrt{J_{2,a}} = \sqrt{\frac{1}{6}\begin{bmatrix} (S_{xx,a} - S_{yy,a})^2 + (S_{yy,a} - S_{zz,a})^2 \\ +(S_{zz,a} - S_{xx,a})^2 + 6(S_{xy,a} - S_{yz,a} - S_{xz,a})^2 \end{bmatrix}}. \qquad [8.23]$$

In this expression, the stress component $S_{ij,a}$ is defined by:

$$S_{ij,a} = \frac{1}{2}\left(\max_{0 \leq t \leq T}[S_{ij}(t)] - min_{0 \leq t \leq T}[S_{ij}(t)]\right);\ i,j = x,y,z. \qquad [8.24]$$

To simplify the calculation of $\sqrt{J_{2,a}}$, this transformation is commonly used [YAI 18]:

$$\sqrt{J_{2,a}} = \sqrt{D_{1,a}^2 + D_{2,a}^2 + D_{3,a}^2 + D_{4,a}^2 + D_{5,a}^2} = \|d_a\| \qquad [8.25]$$

with:

$$\begin{cases} D_1 = \frac{\sqrt{3}}{2}\left(\frac{2}{3}S_{xx} - \frac{1}{3}S_{yy} - \frac{1}{3}S_{zz}\right) \\ D_2 = \frac{1}{2}(S_{yy} - S_{zz}) \\ D_3 = S_{xy} \\ D_4 = S_{yz} \\ D_5 = S_{xz} \end{cases}.$$

The methodology for calculating the average value of $\sqrt{J_{2,a}}$ in the frequency domain is presented in previous works [YAI 18]. As defined by equation [8.22], estimating the semi-axes $R_i$ leads to estimating $\sqrt{J_{2,a}}$. The algorithm for determining $E\left[\sqrt{J_{2,a}}\right]$ is defined as follows:

1) Calculation of the covariance matrix $\Sigma_d$ of $d$, defined by equation [8.26]. It can be determined from the stress covariance matrix $\Sigma_s$ presented by equation [8.17]:

$$\Sigma_d = E\left[\left(d(t) - \bar{d}\right).\left(d(t) - \bar{d}\right)^T\right], \qquad [8.26]$$

$$\Sigma_d = P\Sigma_s P^T. \qquad [8.27]$$

2) The diagonal terms of $\Sigma_d$ present the variance of each component. However, the diagonal terms present the variance between each component. In general, these terms are non-zero. So, $\Sigma_d$ is not diagonal. There is an ortho-normal basis of eigenvectors $A_d$ such as:

$$\Sigma_{\tilde{d}} = A_d \Sigma_d A_d{}^T = diag\left(\sigma_1^2, \sigma_2^2, \sigma_3^2, \sigma_4^2, \sigma_5^2\right), \qquad [8.28]$$

with $\sigma_1 \geq \sigma_2 \geq \sigma_3 \geq \sigma_4 \geq \sigma_5$.

We can then define a vector $\tilde{d}$ associated linearly with $d$ by an orthogonal transformation such as:

$$\tilde{d}(\mathrm{t}) = A_d d(t), \qquad [8.29]$$

with $\tilde{d}(\mathrm{t}) = [\widetilde{D_1}(t)\ \widetilde{D_2}(t)\ \widetilde{D_3}(t)\ \widetilde{D_4}(t)\ \widetilde{D_5}(t)]^T$.

3) Calculation of the number of maxima $N_\mathrm{p}\left(\widetilde{D_\iota}\right)$ of each parameter of the vector $\tilde{d}$. It can be determined from the spectral moment $m_i$ of the random process $\widetilde{D_\iota}$:

$$N_p = \sqrt{\frac{m_4}{m_2}}, \qquad [8.30]$$

$$m_i = \int_{-\infty}^{+\infty} |f|^i \Phi_{\widetilde{D_\iota}}(f) df. \qquad [8.31]$$

$E[R_i]$ can then be estimated using equation [8.31], with $\sigma_{\widetilde{D_\iota}}$ presenting the standard deviation of the random process $\widetilde{D_\iota}$ [YAI 18].

$$E[R_i] \approx \sigma_{\widetilde{D_\iota}}\left(\sqrt{2lnN_{\mathrm{p}}\left(\widetilde{D_\iota}(t)\right)}+\frac{0.5{,}772}{\sqrt{2lnN_{\mathrm{p}}(\widetilde{D_\iota}(t))}}\right), i=1 \ldots 5; \qquad [8.32]$$

4) For structures subjected to random loads, $E\left[\sqrt{J_{2,a}}\right]$ can be determined by Yaich et al. [YAI 18] and Ben Abdessalem et al. [BEN 13]:

$$E\left[\sqrt{J_{2,a}}\right] \approx \sqrt{E[R_1]^2+E[R_2]^2+E[R_3]^2+E[R_4]^2+E[R_5]^2}. \quad [8.33]$$

### 8.6.2. *Fatigue damage*

The implementation of the fatigue damage analysis strategy is shown in Figure 8.13. The algorithm consists of two subproblems: the first subproblem consists of calculating the power spectral densities of the stresses. First, the model must be constructed by defining the geometry, material properties and loads. Then, a modal analysis is necessary to determine the normal modes of the model. We conclude this first subproblem by the spectral analysis to calculate the Power Spectral Densities (PSDs) of the stresses at each element of the structure.

For the second subproblem, and in order to evaluate the fatigue damage, the calculated PSDs are then introduced into the MATLAB software to evaluate the covariance matrix of the stresses in a first step. This allows us to calculate the covariance matrix $\Sigma_d$. The covariance matrix $\Sigma_{\tilde{d}}$ and the PSD $\Phi_{\tilde{d}}$ are obtained by the orthogonal transformation of the matrix $\Sigma_d$. The number of maxima *Np* of each parameter of the vector *d* is calculated. Next, the average of the $E\left[\sqrt{J_{2,a}}\right]$ shear stress amplitude is determined. Finally, the fatigue damage is determined at each point of the structure using the Sines criterion [YAI 18].

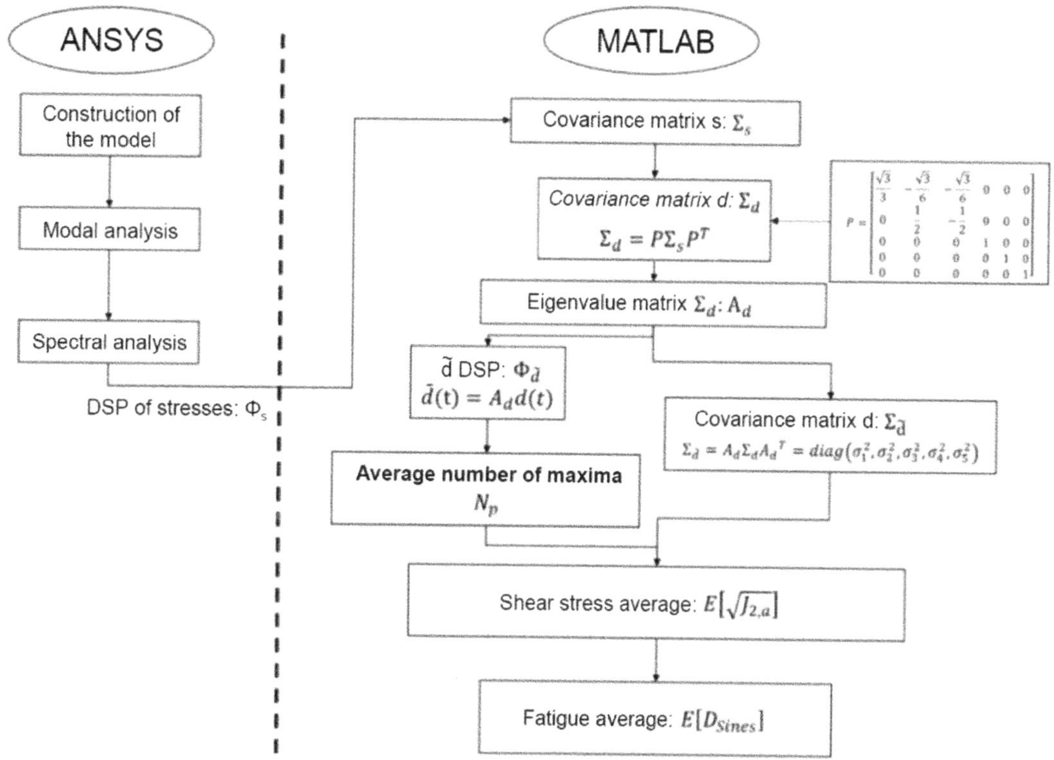


**Figure 8.13.** *Implementation of the fatigue damage analysis strategy*

## 8.7. Methodology of improvement using physical tests

It is necessary to carry out a preliminary study on a product that seeks to improve. For this, a complete and relevant analysis, with the right failure prediction tools and expert advice, is necessary. Once this information has been gathered, it is necessary to consider the nature of future stresses, carry out the dimensioning and develop a mathematical model. Then, we do the tests. It is essential to recover as much data as possible from the behavior of the products and to ensure the adequacy of the parameters during the cycle. From then on, it is possible to check the intended specifications of the designer.

Finally, it is necessary to process the collected data. On the one hand, it is necessary to process the reliability data and, on the other hand, it is essential to identify the causes of failures in order to act on the design.

The product must be corrected or redesigned depending on the type and severity of the failures; for this, all the people involved in the development must be informed of the results in order to choose the corrective actions to be implemented.

NOTE.– Any redesign must be retested to ensure that the changes have not led to other defects with even more serious consequences.

### 8.7.1. *Optimization of the physical tests to be carried out*

#### 8.7.1.1. *Experimental plans*

Since the aggravated tests are extremely expensive, it is necessary first to determine which experiments must be carried out. When performing tests, the behavior of the studied systems depends on large number of parameters, often influencing each other.

The principle of experimental plans is to model a system, to vary these parameters, in order to know the influence on the operation of the overall system.

There are several types of experimental plans:

– Screening plans: these plans allow the most influential parameters to be determined.

– Modeling plans: these plans make it possible to find the input–output law of the system. The best known are the full and fractional factorial designs.

– Mixed plans: these plans are like for modeling plans, but with dependent factors.

If we assume that the studied system can be described by a linear model, we can describe its behavior by a straight line. Two points are then sufficient to obtain its operating mode.

Trials are expensive. More the considered parameters are, more expensive the tests will be. If one wishes to exhaustively study eight variables, each having only two possible values, one will have to carry out $2^8 = 256$ experiments.

This large number of experiments should be reduced to limit costs and investigation time while maintaining the quality and quantity of information from the tests.

The different variables are converted into levels. The lowest value takes level −1 and the highest value takes level +1.

Once the extreme values are calculated, the intermediate values are obtained by linear interpolation between the two extreme points.

The purpose of these experimental plans is to find out a mathematical model of the system function that allows linking the operational response to the input parameters. A polynomial model makes it possible to predict the system behavior, without having to do all the experiments.

Since we are using reduced centered series, we can use the Maclaurin series.

The equation:

$$y = f(x_1, x_2)$$

becomes, considering the second-order Maclaurin series with $i = 1..2$:

$$y = a_0 + \sum a_i x_i + \sum_{i \neq j} a_{ij} x_i x_j + \sum_i a_{ii} x_i x_i,$$

where $y$ is the response, $x_i$ are the parameters and $a_0$, $a_i$, $a_{ij}$, $a_{ii}$ are the coefficients of the mathematical model. They are adjusted using the test results.

Full factorial plan: it is a plan with two factors, two levels, four measures. Obtaining four unknowns and four equations, which allow us to determine the following coefficients: $a_0$, $a_1$, $a_2$, $a_{12}$.

We have an equation of the type:

$$y = a_0 + a_1x_1 + a_2x_2 + a_{12}x_1x_2.$$

It is possible to add an error term $\varepsilon$, but for simplicity we will omit this here.

To solve this equation, it is necessary, in linear behavior, to conduct four experiments in order to arrive at a system of four equations with four unknowns, and thus find each variable.

To do this, we write an "experimentation matrix" where we will find all the possibilities (complete factorial plan).

| Tests | Rest | Factor 1 | Factor 2 | Factor 3 | Answer |
|---|---|---|---|---|---|
| **1** | +1 | –1 | –1 | +1 | $y_1$ |
| **2** | +1 | +1 | –1 | –1 | $y_2$ |
| **3** | +1 | –1 | +1 | –1 | $y_3$ |
| **4** | +1 | +1 | +1 | +1 | $y_4$ |

**Table 8.1.** *Experiment matrix*

We get the following system of equations:

$$y_1 \cong a_0 + a_1x_{1(-1)} + a_2x_{2(-1)} + a_{12}x_{1(-1)}x_{2(-1)}$$
$$y_2 \cong a_0 + a_1x_{1(+1)} + a_2x_{2(-1)} + a_{12}x_{1(+1)}x_{2(-1)}$$
$$y_3 \cong a_0 + a_1x_{1(-1)} + a_2x_{2(+1)} + a_{12}x_{1(-1)}x_{2(+1)}$$
$$y_4 \cong a_0 + a_1x_{1(+1)} + a_2x_{2(+1)} + a_{12}x_{1(+1)}x_{2(+1)}.$$

Thus:

$$\begin{aligned} y_1 &= a_0 - a_1 - a_2 + a_{12} \\ y_2 &= a_0 + a_1 - a_2 - a_{12} \\ y_3 &= a_0 - a_1 + a_2 - a_{12} \\ y_4 &= a_0 + a_1 + a_2 + a_{12} \end{aligned}.$$

We can then calculate the searched unknowns:

$$\begin{aligned} a_0 &= \frac{1}{4}(y_1 + y_2 + y_3 + y_4) \\ a_1 &= \frac{1}{4}(-y_1 + y_2 - y_3 + y_4) \\ a_2 &= \frac{1}{4}(-y_1 - y_2 + y_3 + y_4) \\ a_{12} &= \frac{1}{4}(y_1 - y_2 - y_3 + y_4) \end{aligned}.$$

NOTE.– $a_0$ has a special status, and this is the average theoretical answer.

This calculation can be written and done in matrix form:

$$\begin{bmatrix} a_0 \\ a_1 \\ a_2 \\ a_{12} \end{bmatrix} = \frac{1}{4}\begin{bmatrix} +1 & +1 & +1 & +1 \\ -1 & +1 & -1 & +1 \\ -1 & -1 & +1 & +1 \\ +1 & -1 & -1 & +1 \end{bmatrix} \begin{bmatrix} y_1 \\ y_2 \\ y_3 \\ y_4 \end{bmatrix},$$

which can be written in a general form for second-order linear models:

$$A = \frac{1}{n} XY.$$

The matrix $X$ containing $2^n$ rows is called "full factorial design with interactions".

### 8.7.1.2. *Fractional factorial plans*

Despite the fact that this technique simplifies the modeling of the studied system, it is still necessary to carry out the experiments. The experiments can take a long time and become extremely expensive.

If we had three factors at two levels, it would be necessary to carry out eight experiments. Hence, it may be interesting to carry out fractional experimental plans.

It is possible to obtain a certain amount of information by keeping only certain experiments. Of course, a loss of information is inevitable, but if the interactions are not very strong, this type of experimental design can give quite conclusive results.

EXAMPLE 8.3.–

It is assumed that the service life of a hydraulic motor is expressed in the form:

$$t = KP^a N^b .$$

With $P$ being the oil pressure at the motor inlet and $N$ the rotational speed of the motor shaft, an accelerated test can be carried out by increasing $P$ and $N$ . Let parameters $a$ and $b$ for a motor ($a$ = –3.2, $b$ = –1.09) be given as input data. Under normal operational conditions, the oil pressure is $P1$ = 25 MPa at the speed is $N1$ = 1,500 rpm. While under accelerated conditions, we have $P2$ = 50 MPa at a speed $N2$ = 2,400 rpm.

We can therefore calculate an acceleration factor such as:

$$F_a = \frac{t_1}{t_2} = \left(\frac{P_1}{P_2}\right)^a \times \left(\frac{N_1}{N_2}\right)^b .$$

The acceleration factor, therefore, for this test is:

$$F_a = \frac{t_1}{t_2} = \left(\frac{25}{50}\right)^{-3.2} \times \left(\frac{1,500}{2,400}\right)^{-1.09} = 15.34 .$$

## 8.8. Conclusion

A certain number of tests exist. The empirical models are associated and make it possible to predict the probability of the systems' lifespan. Aggravated tests make it possible to improve the intrinsic quality of the system, from the design stage, by exceeding the real operational conditions. Aggravated testing allows pointing out the latent weaknesses which take years to reveal themselves and thus increasing the statistical lifespan through the correction of the hidden defects. But these tests are expensive to set up to carry out as well as to analyze the experimental results. Carrying out these tests may even turn out to be inconclusive if the wrong experimental choices are made. This is why a preliminary numerical simulation would be useful to explore and save time. Although numerical simulations will never replace real tests, they could help to decide the right choices before incurring significant costs.

# Appendix

## Common Reliability Modeling Laws

There are many distribution laws that model physical observations more or less successfully. All these laws do not represent isolated cases and are of a probabilistic nature. It is necessary to recall that probabilistic figures, in engineering applications, are meaningful if they are accompanied with their associated uncertainty parameters (variance, standard deviation, error interval, confidence level, etc.).

The most cited distributions in system reliability engineering are briefly presented in the following sections.

### A.1. Weibull's law

Weibull's law makes it possible to approximate the three parts of the bathtub curve as a function of the value of the parameter $\beta$ (shape-parameter). Weibull's law is one of the most popular laws in many fields. Because of its flexibility, it can represent infinity of probability distributions. This is the most widely used law, especially in maintenance and electronics.

– Survival function:

$$R(t) = e^{-\left(\frac{t-\gamma}{\eta}\right)^{\beta}}.$$

– Failure rate:

$$\lambda(t) = \frac{\beta}{\eta}\left(\frac{t-\gamma}{\eta}\right)^{\beta-1},$$

with:

– $\eta$ is the scale parameter – this value is independent of time and corresponds to the sum of the operating times of the studied component;

– $\gamma$ is the time-origin parameter;

– $\beta$ is the shape parameter.

1) If $\beta < 1$, the formula models the reliability of the component in its youth period.

2) If $\beta = 1$, it is modeling the useful life period. The component's reliability follows an exponential law and its reliability is independent of its age. It only depends on the duration of the mission, that is, a constant failure rate $\lambda$.

3) If $\beta > 1$, the formula represents the aging phenomena. If $1.5 < \beta < 2.5$, it is for the fatigue part. From $\beta = 3$, the law is assimilated to a normal law and Weibull always remains applicable. If $3 < \beta < 4$, it is for the wear part.

NOTE.– The shape parameter $\beta$ gives indications on the mode of failure and the evolution over time.

## A.2. Exponential law

It is the simplest law. It is used when the studied components have constant failure rates. It is used when equipment fails suddenly and randomly.

– Survival function:

$$R(t) = e^{-\left(\frac{t}{\theta}\right)}.$$

– Failure rate:

$$\lambda(t) = \frac{1}{\theta}.$$

## A.3. Normal law

The normal distribution is a widespread probability distribution and applies to phenomena that are the combination of a large number of random effects.

The density function has two parameters: the mean $\mu$ and the standard deviation $\sigma$:

$$r(t) = \frac{1}{\sigma\sqrt{2\pi}} e^{\left[-\frac{1}{2}\left(\frac{x-\mu}{\sigma}\right)^2\right]}.$$

The distribution function (reliability) is as follows:

$$R(t) = \frac{1}{\sigma\sqrt{2\pi}} \int_{-\infty}^{t} e^{\left[-\frac{1}{2}\left(\frac{x-\mu}{\sigma}\right)^2\right]} dx.$$

## A.4. Lognormal distribution

It can model phenomena that are caused by many factors, none of which is preponderant.

– Survival function (reliability):

$$R(t) = 1 - \Phi\left(\frac{\log(t)-\mu}{\sigma}\right).$$

– Density function:

$$f(t) = \frac{1}{t\sigma\sqrt{2\pi}} e^{\left[-\frac{1}{2}\left(\frac{\log(t)-\mu}{\sigma}\right)^2\right]},$$

$$\lambda(t) = \frac{e^{\left[-\frac{1}{2}\left(\frac{\log(t)-\mu}{\sigma}\right)^2\right]}}{t\int_0^{\infty} \sigma\sqrt{2\pi} f(t) dt}.$$

The lognormal law is often used to represent the distribution of lifetimes of components showing mechanical fatigue.

## A.5. Binomial distribution

It is used to describe events whose repeated tests are independent and identical and have only two possible outcomes, success or failure:

$$f(k|p) = \frac{n!}{k!\,(n-k)}.\ P^{\,k}.(1-p)^{n-k},$$

where $n$ and $k$ represent the total number of repetitive trials and the number of successful observations, respectively, while $p$ is the occurrence probability of the successful observation in only one trial.

## A.6. Poisson's law

It is one of the most widely used distributions for discrete variables. It can be used as an approximation of a binomial distribution when the studied population $n$ is large enough and the occurrence probability $p$ is small. The interest of this distribution is to have only one parameter $\mu$, the average value:

$$P(x) = \frac{\mu^{x}.e^{-\mu}}{x!},$$

with $\mu = n.p$.

## A.7. Gamma law

$$f(t) = \frac{\beta^{\alpha} t^{\alpha-1} e^{-\beta t}}{\Gamma(\alpha)},$$

$$\lambda(t) = \frac{\beta^{\alpha} t^{\alpha-1} e^{-\beta t}}{\int_t^{\infty} \Gamma(\alpha) f(u) du}.$$

## A.8. Beta law

$$F(x) = \frac{\Gamma(a+b)}{\Gamma(a).\Gamma(b)}\ x^{a-1}.(1-x)^{b-1}.$$

## A.9. Law of $\chi^2$

The law of $\chi^2$ is widely used in statistics, for example, to determine confidence intervals, carrying out hypothesis intervals, tests of independence and so forth.

With the distribution function:

$$f(t) = \frac{1}{2^{\frac{v}{2}} \Gamma\left(\frac{v}{2}\right)} t^{\left(\frac{v}{2}-1\right)} e^{\frac{1}{2}}.$$

This law is only valid for a fairly large sample (>30) and as soon as $v$ becomes very large, the law of $\chi^2$ behaves like a reduced centered normal law.

Its expectation $E(\chi_v^2) = v$ and variance $\mathrm{var}(\chi_v^2) = 2v$.

# References


[ABI 20] ABID F., DAMMAK K., EL HAMI A. et al. "Surrogate models for uncertainty analysis of micro-actuator", *Microsystem Technologies*, vol. 26, pp. 2589–2600, 2020.

[ABO 11a] ABO AL-KHEER A., EL-HAMI A., KHARMANDA M.G. et al. "Reliability-based design for soil tillage machines", *Journal of Terramechanics*, vol. 48, no. 1, pp. 57–64, 2011.

[ABO 11b] ABO AL-KHEER A., EID M., AOUES Y. et al. "Theoretical analysis of the spatial variability in tillage forces for fatigue analysis of tillage machines", *Journal of Terramechanics*, vol. 48, no. 4, pp. 285–295, 2011.

[ABO 11c] ABO AL-KHEER A., KHARMANDA M.G., EL HAMI A. et al. "Estimating the variability of tillage forces on a chisel plough shank by modeling the variability of tillage system parameters", *Computers and Electronics in Agriculture*, vol. 78, no. 1, pp. 61–70, 2011.

[BEN 13] BEN ABDESSALEM A., PAGNACCO E., EL HAMI A., "Increasing the stability of T-shape tube hydroforming process under stochastic framework", *International Journal of Advanced Manufacturing Technology*, vol. 69, nos 5–8, pp. 1343–1357, 2013.

[BEN 15] BEN ABDESSALEM A., EL HAMI A., "A probabilistic approach for optimising hydroformed structures using local surrogate models to control failures", *International Journal of Mechanical Sciences*, nos 96–97, pp. 143–162, 2015.

[BOU 96] BOUYEKHF R., EL MOUDNI A., EL HAMI A. et al. "Reduced order modelling of two-time-scale discrete non-linear systems", *Journal of the Franklin Institute*, vol. 333, no. 4, pp. 499–512, 1996.

[BOU 01] BOUYEKHF R., EL HAMI A., EL MOUDNI A., "Optimal control of a particular class of singularly perturbed nonlinear discrete-time systems", *IEEE Transactions on Automatic Control*, vol. 46, no. 7, pp. 1097–1101, 2001.

[CHE 08] CHEROUAT A., RADI B., EL HAMI A., "The frictional contact of the composite fabric's shaping", *Acta Mechanica*, vol. 199, nos 1–4, pp. 29–41, 2008.

[DAM 17] DAMMAK K., EL HAMI A., KOUBAA S. et al. "Reliability based design optimization of coupled acoustic-structure system using generalized polynomial chaos", *International Journal of Mechanical Sciences*, vol. 134, pp. 75–84, 2017.

[DAM 19a] DAMMAK K., EL HAMI A., "Multi-objective reliability-based design optimization of coupled acoustic-structural system", *Engineering Structures*, vol. 197, issue 6, no. 109389, 2019.

[DAM 19b] DAMMAK K., KOUBAA S., EL HAMI A. et al. "Numerical modelling of vibro-acoustic problem in presence of uncertainty: Application to a vehicle cabin", *Applied Acoustics*, vol. 144, pp. 113–123, 2019.

[DAM 20a] DAMMAK K., EL HAMI A., "Multi-objective reliability-based design optimization using Kriging surrogate model for cementless hip prosthesis", *Computer Methods in Biomechanics and Biomedical Engineering*, vol. 23, no. 12, pp. 854–865, 2020.

[DAM 20b] DAMMAK K., YAICH A., EL HAMI A. et al. "An efficient optimization based on the robust hybrid method for the coupled acoustic–structural system", *Mechanics of Advanced Materials and Structures*, vol. 27, no. 21, pp. 1816–1826, 2020.

[DAM 21] DAMMAK K., EL HAMI A., "Thermal reliability-based design optimization using Kriging model of PCM based pin fin heat sink", *International Journal of Heat and Mass Transfer*, vol. 166, no. 120745, 2021.

[DEB 20] DEBICH B., EL HAMI A., YAICH A. et al. "An efficient reliability-based design optimization study for PCM-based heat-sink used for cooling electronic devices", *Mechanics of Advanced Materials and Structures*, vol. 29, issue 12, pp. 1661–1673, 2020.

[DEB 21] DEBICH B., Modélisation, simulation, et fiabilité des matériaux changement de phase pour l'automobile et l'aéronautique, PhD thesis, INSA Rouen Normandie, 2021.

[ELH 93] EL HAMI A., LALLEMENT G., MINOTTI P. et al. "Methods that combine finite group theory with component mode synthesis in the analysis of repetitive structures", *Computers and Structures*, vol. 48, no. 6, pp. 975–982, 1993.

[ELH 96] EL HAMI A., RADI B., "Some decomposition methods in the analysis of repetitive structures", *Computers and Structures*, vol. 58, no. 5, pp. 973–980, 1996.

[ELH 09] EL HAMI A., RADI B., CHEROUAT A., "Treatment of the composite fabric's shaping using a Lagrangian formulation", *Mathematical and Computer Modelling*, vol. 49, nos 7–8, pp. 1337–1349, 2009.

[ELH 11] EL HAMI A., RADI B., "Comparison study of different reliability-based design optimization approaches", *Advanced Materials Research*, vol. 274, pp. 113–121, 2011.

[ELH 13a] EL HAMI A., RADI B., *Incertitude et optimisation et fiabilité des structures*, Hermes-Lavoisier, Paris, 2013.

[ELH 13b] EL HAMI A., RADI B., *Uncertainty and Optimization in Structural Mechanics*, ISTE Ltd, London, and John Wiley & Sons, New York, 2013.

[GUE 15a] GUERINE A., EL HAMI A., WALHA L. et al. "Approach for the dynamic analysis of one stage gear system with uncertain parameters", *Mechanism and Machine Theory*, vol. 92, pp. 113–126, 2015.

[GUE 15b] GUERINE A., EL HAMI A., FAKHFAKH T. et al. "A polynomial chaos method to the analysis of the dynamic behavior of spur gear system", *Structural Engineering and Mechanics*, vol. 53, issue 4, pp. 819–831, 2015.

[GUE 16a] GUERINE A., EL HAMI A., WALHA L. et al. "A polynomial chaos method for the analysis of the dynamic behavior of uncertain gear friction system", *European Journal of Mechanics, A/Solids*, no. 59, pp. 76–84, 2016.

[GUE 16b] GUERINE A., EL HAMI A., WALHA L. et al. "Dynamic response of a Spur gear system with uncertain friction coefficient", *Advances in Engineering Software*, no. 120, pp. 45–54, 2016.

[GUE 17] GUERINE A., EL HAMI A., WALHA L. et al. "Dynamic response of wind turbine gear system with uncertain-but-bounded parameters using interval analysis method", *Renewable Energy*, no. 113, pp. 679–687, 2017.

[HUA 16] HUANG C., RADI B., EL HAMI A., "Uncertainty analysis of deep drawing using surrogate model based probabilistic method", *Manufacturing Technology*, vol. 86, nos 9–12, pp. 3229–3240, 2016.

[JAN 17] JANNOUN M., AOUES Y., PAGNACCO E. et al. "Probabilistic fatigue damage estimation of embedded electronic solder joints under random vibration", *Microelectronics Reliability*, vol. 78, pp. 249–257, 2017.

[KAM 20] KAMEL A., DAMMAK K., EL HAMI A. et al. "A modified hybrid method for a reliability-based design optimization applied to an offshore wind turbine", *Mechanics of Advanced Materials and Structures*, vol. 29, 2022 – issue 9, pp. 1229–1242, 2020.

[KHA 08] KHARMANDA G., MOHSINE A., MAKLOUFI A. et al. "Recent methodologies for reliability-based design optimization", *International Journal for Simulation and Multidisciplinary Design Optimization*, vol. 2, no. 1, pp. 11–23, 2008.

[KHA 09] KHARMANDA G., SHARABATEY S., IBRAHIM H. et al. "Reliability-based design optimization using semi-numerical strategies for structural engineering applications", *International Journal of CAD/CAM 9*, vol. 9, issue 1, pp. 1–16, 2009.

[LED 17a] LEDOUX M., EL HAMI A., *Fluid Mechanics*, ISTE Ltd, London, and John Wiley & Sons, New York, 2017.

[LED 17b] LEDOUX M., EL HAMI A., *Compressible Flow Propulsion and Digital Approaches in Fluid Mechanics*, ISTE Ltd, London, and John Wiley & Sons, New York, 2017.

[LED 22a] LEDOUX M., EL HAMI A., *Heat Transfer 1: Conduction*, ISTE Ltd, London, and John Wiley & Sons, New York, 2022.

[LED 22b] LEDOUX M., EL HAMI A., *Heat Transfer 2: Radiative Transfer*, ISTE Ltd, London, and John Wiley & Sons, New York, 2022.

[MAB 17] MABROUK I.B., EL HAMI A., WALHA L. et al. "Dynamic vibrations in wind energy systems: Application to vertical axis wind turbine", *Mechanical Systems and Signal Processing*, vol. 85, pp. 396–414, 2017.

[MAK 16] MAKHLOUFI A., AOUES Y., EL HAMI A., "Reliability based design optimization of wire bonding in power microelectronic devices", *Microsystem Technologies*, vol. 22, no. 12, pp. 2737–2748, 2016.

[MOH 10] MOHSINE A., EL HAMI A., "A robust study of reliability-based optimization methods under eigen-frequency", *Computer Methods in Applied Mechanics and Engineering*, vol. 199, nos 17–20, pp. 1006–1018, 2010.

[MOR 02] MORO T., EL HAMI A., EL MOUDNI A., "Reliability analysis of a mechanical contact between deformable solids", *Probabilistic Engineering Mechanics*, vol. 17, no. 3, pp. 227–232, 2002.

[RAD 07] RADI B., EL HAMI A., "Reliability analysis of the metal forming process", *Mathematical and Computer Modelling*, vol. 45, nos 3–4, pp. 431–439, 2007.

[SAR 11] SARSRI D., AZRAR L., JEBBOURI A. et al. "Component mode synthesis and polynomial chaos expansions for stochastic frequency functions of large linear FE models", *Computers and Structures*, vol. 89, nos 3–4, pp. 346–356, 2011.

[YAI 18] YAICH A., KHARMANDA G., EL HAMI A. et al. "Reliability based design optimization for multiaxial fatigue damage analysis using robust hybrid method", *Journal of Mechanics*, vol. 34, no. 5, pp. 551–566, 2018.

# Index

## T, U, V

Other titles from

ISTE

in

Mechanical Engineering and Solid Mechanics

## 2023

EL HAMI Abdelkhalak, DELAUX David, GRZESKOWIAK Henri
*Applied Reliability for Industry 1: Predictive Reliability for the Automobile, Aeronautics, Defense, Medical, Marine and Space Industries*
*(Reliability of Multiphysical Systems Set – Volume 16)*
*Applied Reliability for Industry 2: Experimental Reliability for the Automobile, Aeronautics, Defense, Medical, Marine and Space Industries*
*(Reliability of Multiphysical Systems Set – Volume 17)*
*Applied Reliability for Industry 3: Operational Reliability for the Automobile, Aeronautics, Defense, Medical, Marine and Space Industries*
*(Reliability of Multiphysical Systems Set – Volume 18)*

## 2022

BAYLE Franck
*Product Maturity 1: Theoretical Principles and Industrial Applications*
*(Reliability of Multiphysical Systems Set – Volume 12)*
*Product Maturity 2: Principles and Illustrations*
*(Reliability of Multiphysical Systems Set – Volume 13)*

EL HAMI Abdelkhalak, DELAUX David, GRZESKOWIAK Henri
*Reliability and Physics-of-Healthy in Mechatronics*
*(Reliability of Multiphysical Systems Set – Volume 15)*

LANNOY André
*Reliability of Nuclear Power Plants: Methods, Data and Applications*
*(Reliability of Multiphysical Systems Set – Volume 14)*

LEDOUX Michel, EL HAMI Abdelkhalak
*Heat Transfer 3: Convection, Fundamentals and Monophasic Flows*
*(Mathematical and Mechanical Engineering Set – Volume 11)*
*Heat Transfer 4: Convection, Two-Phase Flows and Special Problems*
*(Mathematical and Mechanical Engineering Set – Volume 12)*

PLANCHETTE Guy
*Cindynics, The Science of Danger: A Wake-up Call*
*(Reliability of Multiphysical Systems Set – Volume 11)*

## 2021

CHALLAMEL Noël, KAPLUNOV Julius, TAKEWAKI Izuru
*Modern Trends in Structural and Solid Mechanics 1: Static and Stability*
*Modern Trends in Structural and Solid Mechanics 2: Vibrations*
*Modern Trends in Structural and Solid Mechanics 3: Non-deterministic Mechanics*

DAHOO Pierre Richard, POUGNET Philippe, EL HAMI Abdelkhalak
*Applications and Metrology at Nanometer Scale 1: Smart Materials, Electromagnetic Waves and Uncertainties*
*(Reliability of Multiphysical Systems Set – Volume 9)*
*Applications and Metrology at Nanometer Scale 2: Measurement Systems, Quantum Engineering and RBDO Method*
*(Reliability of Multiphysical Systems Set – Volume 10)*

LEDOUX Michel, EL HAMI Abdelkhalak
*Heat Transfer 1: Conduction*
*(Mathematical and Mechanical Engineering Set – Volume 9)*
*Heat Transfer 2: Radiative Transfer*
*(Mathematical and Mechanical Engineering Set – Volume 10)*

## 2020

SALENÇON Jean
*Elastoplastic Modeling*

## 2019

BAYLE Franck
*Reliability of Maintained Systems Subjected to Wear Failure Mechanisms: Theory and Applications*
*(Reliability of Multiphysical Systems Set – Volume 8)*

BEN KAHLA Rabeb, BARKAOUI Abdelwahed, MERZOUKI Tarek
*Finite Element Method and Medical Imaging Techniques in Bone Biomechanics*
*(Mathematical and Mechanical Engineering Set – Volume 8)*

IONESCU Ioan R., QUEYREAU Sylvain, PICU Catalin R., SALMAN Oguz Umut
*Mechanics and Physics of Solids at Micro- and Nano-Scales*

LE VAN Anh, BOUZIDI Rabah
*Lagrangian Mechanics: An Advanced Analytical Approach*

MICHELITSCH Thomas, PÉREZ RIASCOS Alejandro, COLLET Bernard, NOWAKOWSKI Andrzej, NICOLLEAU Franck
*Fractional Dynamics on Networks and Lattices*

SALENÇON Jean
*Viscoelastic Modeling for Structural Analysis*

VÉNIZÉLOS Georges, EL HAMI Abdelkhalak
Movement Equations 5: Dynamics of a Set of Solids
*(Non-deformable Solid Mechanics Set – Volume 5)*

## 2018

BOREL Michel, VÉNIZÉLOS Georges
*Movement Equations 4: Equilibriums and Small Movements*
*(Non-deformable Solid Mechanics Set – Volume 4)*

FROSSARD Etienne
*Granular Geomaterials Dissipative Mechanics: Theory and Applications in Civil Engineering*

RADI Bouchaib, EL HAMI Abdelkhalak
*Advanced Numerical Methods with Matlab® 1: Function Approximation and System Resolution*
*(Mathematical and Mechanical Engineering Set – Volume 6)*
*Advanced Numerical Methods with Matlab® 2: Resolution of Nonlinear, Differential and Partial Differential Equations*
*(Mathematical and Mechanical Engineering Set – Volume 7)*

SALENÇON Jean
*Virtual Work Approach to Mechanical Modeling*

## 2017

BOREL Michel, VÉNIZÉLOS Georges
*Movement Equations 2: Mathematical and Methodological Supplements*
*(Non-deformable Solid Mechanics Set – Volume 2)*
*Movement Equations 3: Dynamics and Fundamental Principle*
*(Non-deformable Solid Mechanics Set – Volume 3)*

BOUVET Christophe
*Mechanics of Aeronautical Solids, Materials and Structures*
*Mechanics of Aeronautical Composite Materials*

BRANCHERIE Delphine, FEISSEL Pierre, BOUVIER Salima, IBRAHIMBEGOVIC Adnan
*From Microstructure Investigations to Multiscale Modeling: Bridging the Gap*

CHEBEL-MORELLO Brigitte, NICOD Jean-Marc, VARNIER Christophe
*From Prognostics and Health Systems Management to Predictive Maintenance 2: Knowledge, Traceability and Decision*
*(Reliability of Multiphysical Systems Set – Volume 7)*

EL HAMI Abdelkhalak, RADI Bouchaib
*Dynamics of Large Structures and Inverse Problems*
*(Mathematical and Mechanical Engineering Set – Volume 5)*
*Fluid-Structure Interactions and Uncertainties: Ansys and Fluent Tools*
*(Reliability of Multiphysical Systems Set – Volume 6)*

KHARMANDA Ghias, EL HAMI Abdelkhalak
*Biomechanics: Optimization, Uncertainties and Reliability*
*(Reliability of Multiphysical Systems Set – Volume 5)*

LEDOUX Michel, EL HAMI Abdelkhalak
*Compressible Flow Propulsion and Digital Approaches in Fluid Mechanics*
*(Mathematical and Mechanical Engineering Set – Volume 4)*
*Fluid Mechanics: Analytical Methods*
*(Mathematical and Mechanical Engineering Set – Volume 3)*

MORI Yvon
*Mechanical Vibrations: Applications to Equipment*

## 2016

BOREL Michel, VÉNIZÉLOS Georges
*Movement Equations 1: Location, Kinematics and Kinetics*
*(Non-deformable Solid Mechanics Set – Volume 1)*

BOYARD Nicolas
*Heat Transfer in Polymer Composite Materials*

CARDON Alain, ITMI Mhamed
*New Autonomous Systems*
*(Reliability of Multiphysical Systems Set – Volume 1)*

DAHOO Pierre Richard, POUGNET Philippe, EL HAMI Abdelkhalak
*Nanometer-scale Defect Detection Using Polarized Light*
*(Reliability of Multiphysical Systems Set – Volume 2)*

DE SAXCÉ Géry, VALLÉE Claude
*Galilean Mechanics and Thermodynamics of Continua*

DORMIEUX Luc, KONDO Djimédo
*Micromechanics of Fracture and Damage*
*(Micromechanics Set – Volume 1)*

EL HAMI Abdelkhalak, RADI Bouchaib
*Stochastic Dynamics of Structures*
*(Mathematical and Mechanical Engineering Set – Volume 2)*

GOURIVEAU Rafael, MEDJAHER Kamal, ZERHOUNI Noureddine
*From Prognostics and Health Systems Management to Predictive Maintenance 1: Monitoring and Prognostics*
*(Reliability of Multiphysical Systems Set – Volume 4)*

KHARMANDA Ghias, EL HAMI Abdelkhalak
*Reliability in Biomechanics*
*(Reliability of Multiphysical Systems Set –Volume 3)*

MOLIMARD Jérôme
*Experimental Mechanics of Solids and Structures*

RADI Bouchaib, EL HAMI Abdelkhalak
*Material Forming Processes: Simulation, Drawing, Hydroforming and Additive Manufacturing*
*(Mathematical and Mechanical Engineering Set – Volume 1)*

## 2015

KARLIČIĆ Danilo, MURMU Tony, ADHIKARI Sondipon, MCCARTHY Michael
*Non-local Structural Mechanics*

SAB Karam, LEBÉE Arthur
*Homogenization of Heterogeneous Thin and Thick Plates*

## 2014

ATANACKOVIC M. Teodor, PILIPOVIC Stevan, STANKOVIC Bogoljub, ZORICA Dusan
*Fractional Calculus with Applications in Mechanics: Vibrations and Diffusion Processes*
*Fractional Calculus with Applications in Mechanics: Wave Propagation, Impact and Variational Principles*

CIBLAC Thierry, MOREL Jean-Claude
*Sustainable Masonry: Stability and Behavior of Structures*

ILANKO Sinniah, MONTERRUBIO Luis E., MOCHIDA Yusuke
*The Rayleigh–Ritz Method for Structural Analysis*

LALANNE Christian
*Mechanical Vibration and Shock Analysis – 5-volume series – 3rd edition*
*Sinusoidal Vibration – Volume 1*
*Mechanical Shock – Volume 2*
*Random Vibration – Volume 3*
*Fatigue Damage – Volume 4*
*Specification Development – Volume 5*

LEMAIRE Maurice
*Uncertainty and Mechanics*

## 2013

ADHIKARI Sondipon
*Structural Dynamic Analysis with Generalized Damping Models: Analysis*

ADHIKARI Sondipon
*Structural Dynamic Analysis with Generalized Damping Models: Identification*

BAILLY Patrice
*Materials and Structures under Shock and Impact*

BASTIEN Jérôme, BERNARDIN Frédéric, LAMARQUE Claude-Henri
*Non-smooth Deterministic or Stochastic Discrete Dynamical Systems: Applications to Models with Friction or Impact*

EL HAMI Abdelkhalak, RADI Bouchaib
*Uncertainty and Optimization in Structural Mechanics*

KIRILLOV Oleg N., PELINOVSKY Dmitry E.
*Nonlinear Physical Systems: Spectral Analysis, Stability and Bifurcations*

LUONGO Angelo, ZULLI Daniele
*Mathematical Models of Beams and Cables*

SALENÇON Jean
*Yield Design*

## 2012

DAVIM J. Paulo
*Mechanical Engineering Education*

DUPEUX Michel, BRACCINI Muriel
*Mechanics of Solid Interfaces*

ELISHAKOFF Isaac *et al.*
*Carbon Nanotubes and Nanosensors: Vibration, Buckling and Ballistic Impact*

GRÉDIAC Michel, HILD François
*Full-Field Measurements and Identification in Solid Mechanics*

GROUS Ammar
*Fracture Mechanics* – 3-volume series
*Analysis of Reliability and Quality Control – Volume 1*
*Applied Reliability – Volume 2*
*Applied Quality Control – Volume 3*

RECHO Naman
*Fracture Mechanics and Crack Growth*

## 2011

KRYSINSKI Tomasz, MALBURET François
*Mechanical Instability*

SOUSTELLE Michel
*An Introduction to Chemical Kinetics*

## 2010

BREITKOPF Piotr, FILOMENO COELHO Rajan
*Multidisciplinary Design Optimization in Computational Mechanics*

DAVIM J. Paulo
*Biotribology*

PAULTRE Patrick
*Dynamics of Structures*

SOUSTELLE Michel
*Handbook of Heterogenous Kinetics*

## 2009

BERLIOZ Alain, TROMPETTE Philippe
*Solid Mechanics using the Finite Element Method*

LEMAIRE Maurice
*Structural Reliability*

## 2007

GIRARD Alain, ROY Nicolas
*Structural Dynamics in Industry*

GUINEBRETIÈRE René
*X-ray Diffraction by Polycrystalline Materials*

KRYSINSKI Tomasz, MALBURET François
*Mechanical Vibrations*

KUNDU Tribikram
*Advanced Ultrasonic Methods for Material and Structure Inspection*

SIH George C. *et al.*
*Particle and Continuum Aspects of Mesomechanics*

Printed and bound by CPI Group (UK) Ltd, Croydon, CR0 4YY

13/09/2026

14961042-0003